Respiratory Physiology

T0180166

Respiratory Physiology
Understanding Gas Exchange

Henry D. Prange

Associate Professor of Physiology and Biophysics
Medical Sciences Program
at Indiana University in Bloomington

CHAPMAN & HALL

New York • Albany • Bonn • Boston • Cincinnati • Detroit • London • Madrid • Melbourne
Mexico City • Pacific Grove • Paris • San Francisco • Singapore • Tokyo • Toronto • Washington

1 2 3 4 5 6 7 8 9 10 XXX 01 00 99 98 97 96 95

Library of Congress Cataloging-in-Publication Data

Prange, Henry D., 1942-
    Respiratory physiology : undrstanding gas exchange / by Henry D.
Prange.
      p.   cm.
    Includes bibliographical references and index.
    ISBN 0-412-05211-3 (alk. paper)
    1. Pulmonary gas exchange. 2. Respiration.   I. Title.
    [DNLM: 1. Respiratory Transport--physiology.   WF 102 P899r 1995]
    QP124. P73 1995
    612. 2'2--dc20
    DNLM/DLC
    for Library of Congress   95-24639
                                     CIP

British Library Cataloguing in Publication Data available

With thanks to my mentor, Knut Schmidt-Nielsen

CONTENTS

PREFACE

Why write another small book on respiratory physiology? I have a dozen or so texts on my bookshelf that could already be used interchangeably to teach the subject. For profit, I might as well buy lottery tickets. Not that my publisher is ungenerous, you understand, it's just that the market is not that big and there are many contenders for a share. No, I write from the idealistic standpoint that I think I have something different to say, something that is importantly different about how gas exchange works and with an approach that is different from other authors.

With few changes, basically the same text or chapters on respiratory physiology have been written, by different authors, for decades. One could almost interchange the tables of contents of most of them. Most seem to have copied the figures and concepts used by the others. Few have done more than accept and perpetuate the conventional wisdom.

In this text, I have attempted to start from fundamental principles of biology, chemistry, and physics and ask at each step, "Does it make sense?" The mechanisms and structures of gas exchange exist because, scientifically and logically, they "can't not be" as they are. The nature of our environment and the capabilities of living tissue are such that only certain opportunities have been available to the evolution of gas exchange.

I have tried to go to the original references rather than other texts for my source material. (In doing so, the German I once learned has again become useful and, in many cases, I found that the original authors have often been misquoted or misinterpreted.) It has been a rewarding learning experience. Even if this text were never to be published, I have already gathered more than sufficient reward from what I have learned to justify the labor. (See, I told you I wasn't in it for the money).

I believe in the approach I have taken and that it is better than the conventional approach. Even students who have hated it (and me) have come back to say they found later that they really understood what was going on (of course, they never said if they liked me any better because of their late insight, but I would rather be appreciated than liked if I have to choose).

I believe that understanding the fundamental processes demystifies physiology. It also allows one to synthesize and explore relationships among what may seem to be unrelated functions if they are merely described. I also believe in the motto, adapted freely from Descartes, "Dubito, ergo sum", I doubt, therefore I am. I believe that skepticism is a critical skill in science. We must recognize , as Nietzsche said, that "Seeing and not believing is the scientist's greatest talent and appearance is his (or her) greatest enemy."

Throughout this book I have espoused new and unconventional explanations for the mechanisms of respiratory physiology. I do this not just because I am enamored of the novel and sensational but because I think they offer a better understanding of the nature of the subject. When the conventional wisdom doesn't make sense, it is the scientist's duty to reexamine the subject. I am sure I have advocated ideas in this book, both novel and conventional, that, sooner or later, will be shown to be wrong, in part or in sum. Where readers will point out my errors, they will have done us all a favor: We all will learn.

So, reader, I commend my book to you in that I hope that, through it, you will learn. Whether by being convinced by my arguments or by refuting them, it matters not. That is what it's all about.

ACKNOWLEDGMENTS

I wrote this book by myself and I willingly accept the responsibility for whatever liability comes from it. Yet, in many ways, my contribution has been only a small part of the finished work. I could not have written it without the encouragement, inspiration, and education I have received from others. It has been my good fortune to know and learn from some of the greatest scientists of our era. What stature my work may have derives primarily from theirs, upon which mine is founded.

I have had the further good fortune to spend my career in the company some of the best students and colleagues one could imagine. To try to list them all would risk inadvertent omissions and lengthen this section of the text more than the publisher would allow. Their continued and generous support has been an essential ingredient to my work.

This book had its origins as a series of lectures I gave as Dozor Visiting Professor at Ben-Gurion University of the Negev.

My thanks must also go to Greg Payne, my editor and friend. He has patiently encouraged this endeavor for longer than I care to say.

CHAPTER 1

GROUND RULES OF GAS EXCHANGE

Gas exchange is the fundamental process of respiration. I include any movement of respiratory gases anywhere between the environment and the membranes of the mitochondria in the definition. For the purposes of this text, the terms *respiration* and *gas exchange* are synonymous. Although I will deal almost exclusively with the gas exchange of mammals and, within that group, primarily with humans, the principles and concepts I discuss pertain just as well to the other species.

For most of the organisms on earth, oxygen is the abundant, available and ideal acceptor of electrons generated in the breakdown of metabolic fuels. The addition of photosynthetic oxygen to the Precambrian atmosphere more than 2 million years ago (Budyko et al., 1987) set in motion an inevitable chain of events in organisms. In contrast to other electron acceptors then present, oxygen was not just more active chemically; its existence as a gas meant that it was more uniformly distributed and more easily transported. The consequent increased potential for oxidation enhanced substantially the energetic drive of evolution: Life became possible in more diverse places and in more forms than ever before.

The by-products of oxidation are carbon dioxide and water.

CO_2 is a distinctly different gas from O_2. It is much more soluble in water, it more readily permeates biological membranes, it participates in a weak acid dissociation, and its exchange may vastly influence the acid-base status of the organism. Thus, carbon dioxide, as well as oxygen, critically affects the respiratory functions and structures of organisms. Other gases involved in respiratory gas exchange must also be considered. These are the gases, chiefly nitrogen and argon, that we lump together as physiologically inert, as well as water vapor. Although, when compared with oxygen and carbon dioxide, they may seem to be minor players, nevertheless their roles in respiration are not inconsequential.

Organisms are not exempt from the rules of physics and chemistry. They are constrained in their evolution of gas exchangers, development of respiratory strategies and exploitation of environments by those characteristics peculiar to oxygen, carbon dioxide, air, water, and the other constituents of gas exchange.

Respiratory gases do not exist in isolation. As they pass through the stages of gas exchange, each gas interacts with the various media in which they move. Therefore, to understand the nature of gas exchange, it is necessary to understand the nature of its components. Fortunately, the traits of the respiratory gases and their interactions with their media are relatively simple and easily learned. Armed with this set of *ground rules* one can examine and understand the bases of respiratory gas exchange.

Quantities of Gases

In a biological context, quantities of gases are typically given as volumes corrected to standard temperature and pressure (STP). This format, although chemically and physically correct, is troublesome for biological purposes. The correction to STP provides a wondrous opportunity for errors in conversion and application. Use of quantities related to mass (grams or moles) rather than volume, although less common, is far less prone to generation of errors because no temperature or pressure correction is required. In units appropriate to the magnitude of most biological exchange, quantities of gases should be in milligram (mg) or millimole (mM).

Volume

Air as a medium is appropriately quantified as a volume at the conditions prevailing when and where the organism is carrying on its gas exchange. Any "correction" to STP would render the volume irrelevant to the organism (another good reason to avoid that format). Commonly, volumes of air will be in terms of liters or cubed linear dimensions such as cubic centimeters with temperature either stated or safely presumed.

Volumes of water or other biological liquids need not be corrected. The compressibility of water is not important to gas exchange except, perhaps, at abyssal depths of the ocean. Temperature and the presence of solutes are critical to the interaction of water with respiratory gases but not to its volume.

Concentration

Respiratory gases are soluble in water, but their solubility varies greatly, depending on the gas pressure and the particular species of gas. To be consistent, the concentrations of dissolved respiratory gases should be given in units of mass/volume, e.g., mg/L or mM/L. Commonly, this ideal is ignored. Gas concentrations are most frequently given as volumes of gas (STP) per volume of liquid. More confusing still is that, in physiological contexts, the volume of the liquid is often reported as a deciliter (100 ml). Maneuvering through this maze of units requires close attention and a good system for conversion. In Text Box 1-1, I have given an example of what is, in my experience, the best system for conversions of units.

In respiratory physiology, concentration includes the fractional components of gases in gas mixtures and the amounts

---

Text Box 1-1.   Conversions of Units

Any quantity may be multiplied by a fraction without changing its value if the numerator and denominator of the fraction are equivalents, that is, the fraction equals 1. Conversion factors written as fractions that are mathematically equivalent to 1 allow an offending unit to be canceled algebraically and replaced by a preferred unit. After the cancellation, the numerical coefficient in the fraction will remain in the correct position to convert the value of the term.

For example, to convert an oxygen concentration given as an STP volume of 20.1 ml/dl of blood to standard SI units, note that 1 mM of gas is equivalent to 22.4 ml (STP), a mM of O_2 has an equivalent mass of 32 mg and that there are 10 dl in a liter. Using the above equivalents as fractions that equal 1 produces the following equation:

$$\frac{20.1 \text{ ml } O_2}{1 \text{ dl blood}} \; \frac{1 \text{ mM}}{22.4 \text{ ml}} \; \frac{32 \text{ mg}}{1 \text{ mM}} \; \frac{10 \text{ dl}}{1 \text{ L}} = \frac{287 \text{ mg } O_2}{1 \text{ L blood}}$$

The unwanted dimensions will cancel only if the conversion factors are in the correct position. Because each of the factors has a value of 1, any may be inverted to suit the convenience of the user. Examples of some other useful conversion factors in this format are in Table 1-1.

---

carried in volumes of other, more complex media such as blood. In both of these cases, the dimensions are the same as in aqueous solutions.

Partial Pressure

The concept of the partial pressure of a gas is often a source of confusion. For a given gas in a gas mixture, it can be simply and rigorously defined as that part of the total pressure of a gas mixture that can be attributed to the gas in question. That part of the

Table 1-1.   Conversion Factors

Units of Mass and Volume	Units of Force and Pressure
$\dfrac{mM}{22.4\,ml\,(STP)} = 1.0$	$\dfrac{kPa}{7.5\,mmHg} = 1.0$
$\dfrac{liter}{0.0353\,ft^3} = 1.0$	$\dfrac{mmHg}{0.131\,kPa} = 1.0$
Units of Work, Energy and Power	$\dfrac{mbar}{0.1\,kPa} = 1.0$
$\dfrac{kcal\,h^{-1}}{1.16\,W} = 1.0$	$\dfrac{cm\,H_2O}{0.098\,kPa} = 1.0$
$\dfrac{kcal}{4.184\,kJ} = 1.0$	$\dfrac{kPa}{9.87 \times 10^{-3}\,atm} = 1.0$
$\dfrac{kcal\,day^{-1}}{0.0484\,W} = 1.0$	
$\dfrac{kW}{0.239\,kcal\,s^{-1}} = 1.0$	

Note: The following quantities are arranged as fractions that have the value of 1.0. They may be inverted as required to facilitate cancellation.

total pressure is directly proportional to the fraction of the gas mixture that the gas comprises. This definition is illustrated in Fig. 1-1.

But now the problems begin. The dimensions of pressure are force/area, but they are inconveniently concealed in the several units conventionally used to express partial pressure. Nothing in the terms "Pascal," "Torr," "atmosphere," or "mmHg" gives any clue to the physical definition of pressure. Add to this the use of the term "tension" to mean partial pressure, the idea of the vapor pressure of water, and the notion of the partial pressure of a gas dissolved in a liquid, and confusion is all but guaranteed.

As I will discuss later, partial pressure can be thought of as the potential for diffusive exchange. At equilibrium, the partial pressure of a gas dissolved in water is, by definition, the same as its partial pressure in the gas mixture associated with it. For example (Fig. 1-2), a body of water that is well stirred will have the same partial pressure of oxygen and other gases as the air above.

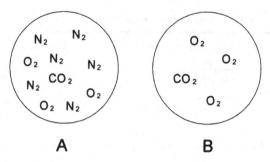

A   **B**

Fig. 1-1. The partial pressure of a gas (P_{gas}) in a mixture is defined as the product of the total pressure (P_{total}) of the mixture and the fraction (F_{gas}) of the mixture that gas comprises. Thus $P_{gas} = P_{total} \times F_{gas}$. To illustrate this concept, assume that in container A there is a gas mixture of 60% N_2, 30% O_2 and 10% CO_2 at a total pressure of 100 kPa. The partial pressure of oxygen (P_{O_2}) is 0.3×100 kPa $= 30$ kPa. Similarly, the P_{N_2} is 60 kPa and the P_{CO_2} is 10 kPa. In container B the mixture is 75% O_2 and 25% CO_2 at a total pressure of 40 kPa. The P_{O_2} and P_{CO_2} are again 30 kPa and 10 kPa, respectively. The P_{N_2} in "B" is, of course, zero. Despite the differences in total pressure and gas composition in the two containers, the potential for diffusion from the container for O_2, for example, is the same in each container because the partial pressure of oxygen is the same in each.

This equilibrium occurs because the net diffusional exchange of a gas between the air and water becomes zero when the partial pressures of the gas in each medium are the same.

It is most important to note that the partial pressure of a gas in a medium, by itself, does not allow one to conclude anything about the concentration of that gas. Partial pressure alone describes only the potential for diffusion. It is necessary to have more specific information about the gas and the medium before one can infer concentration.

Capacitance

The partial pressure of a gas dissolved in a liquid and the concentration of that gas are related by a solubility coefficient that is specific to temperature, the particular gas and the liquid. The

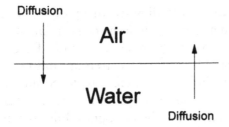

Fig. 1-2. Net diffusion occurs between media as a function of the partial pressure difference between them. At equilibrium, the partial pressures in each medium will be the same. Although random movement of molecules continues between the media even at equilibrium, the net transport between them will be zero. It is important to note that, although factors like capacitance and concentration may affect the rate of transfer by diffusion, *only* partial pressure difference determines whether or not net diffusion will occur.

product of the partial pressure and this coefficient equals the concentration. The solubility of a gas in a liquid decreases as the temperature of the solution increases. This latter phenomenon is easily observed by warming a carbonated beverage and noting the formation of gas bubbles.

One can find, in reference texts, tables of solubility coefficients. These numbers, in units of ml_{gas} (STP)/(760 mmHg×L), are called Bunsen solubility coefficients and are given the symbol α. Although these units are somewhat cumbersome, because we do not often encounter gases at a partial pressure of 760 mmHg, one can see from the dimensions of α that the amount of a gas dissolved is a function of its partial pressure and the solubility coefficient.

The use of α is satisfactory for simple solutions. In the context of physiological gas exchange, however, one must deal with circumstances in which the concept of solubility is inappropriate, such as the complicated associations of gases and blood and the carriage of a gas in a gaseous medium.

To allow more facile comparison of physiological gas transport and exchange between media, Piiper et al. (1971) created the concept of capacitance. They gave capacitance the symbol β and defined it in a more general way than solubility

(β = ΔConcentration/ΔPartial pressure) without concern for the particular medium in or mechanism by which a gas was carried. They also gave β more useful units for concentration, μM/L, and partial pressure was given in units of Torr or mmHg rather than a whole atmosphere. I prefer SI units for β, μM/(L×kPa).

For simple aqueous solutions, α and β are identical. They differ only in the choice of units. For gases in a gaseous medium, β can be defined in terms of the absolute temperture and the gas constant so that β = 1/(RT) (Piiper et al. 1971); it has the same value for all gases that behave reasonably closely to ideality. For carriage in complex media such as blood, where the amount of gas does not vary linearly with partial pressure, β is the difference in concentration that occurs between the partial pressures that exist before and after exchange has occurred. In the case of blood as it goes through gas exchange in the lungs,

$$\beta_{blood_{O_2}} = \frac{C_{a_{O_2}} - C_{\bar{v}_{O_2}}}{P_{a_{O_2}} - P_{\bar{v}_{O_2}}}$$

The symbol C refers to concentration, P to partial pressure, a to arterial blood, and $\bar{v}$ to mixed venous blood. The relationship of concentration to partial pressure for blood is distinctly nonlinear. The definition of β in this case deals with the difference in the initial and final conditions in a given circumstance without regard for the intervening changes. Graphically, it is the slope of the line relating concentration on the vertical axis to partial pressure on the horizontal axis.

$$\text{Capacitance} = \frac{\text{Concentration}}{\text{Partial Pressure}}$$

In the case of blood, β will change as different combinations of partial pressure and concentration are encountered. The solubility coefficient, α, is a constant at all combinations and cannot be correctly used to describe the carriage of oxygen by the blood. β replaces solubility in simple solutions and also applies to gas mixtures. Whatever the medium, the same term applies and it still has the same definition.

To describe gas exchange one needs to deal with the amounts of gas that move across a gas exchanger. The benefit of capacitance is that it allows us to follow gas exchange through the various media of respiration without changing units. For example, with knowledge of capacitance, one can compare the amounts of blood and air that need to pass an exchange surface because the change in partial pressure will be the same for both media. A comparison of values for β among commonly encountered media is illustrated in Figs. 1-3a and 1-3b. The concept of capacitance will be found repeatedly as the modes of gas exchange are examined. Capacitance concisely describes the combined properties of a gas and its medium; it is, simply, the most important fundamental idea in the understanding of gas exchange.

Gas Transport

The first fundamental concept of gas exchange, capacitance, describes the combination of gases and their media. The two remaining concepts, diffusion and bulk flow, deal with the means by which gases move from one site to another. From these concepts onward, the understanding of respiration will, of course, become more complex, but everything thereafter is derived from these three concepts.

Diffusion

Distinct from the physical pressure on the medium, the random movement of gas molecules within it give them a potential for movement that we refer to as the partial pressure. The average

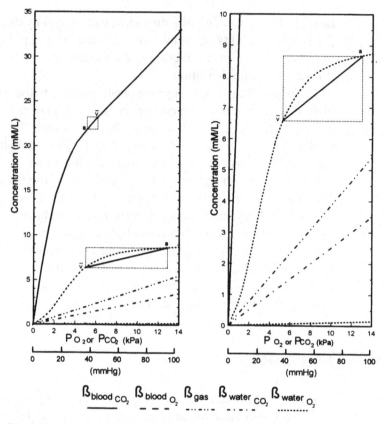

Fig. 1-3. Capacitance (β) is the slope of the line that results when concentration is plotted against partial pressure for a given gas and respiratory medium. The differences in the curves for air, water, and human blood show the differences in β that gas exchange encounters. In the left graph, the scale is large enough to demonstrate the entire range of capacitances. The portions of the curves for O_2 and CO_2 that are enclosed show the portions of the curve that might typically be involved in gas exchange. Adjacent to the boxes, the *a* is the arterial concentration and partial pressure; the $\bar{v}$ is the mixed venous concentration and partial pressure. Note that the changes in concentration are nearly identical for both gases but that, because of its much larger capacitance, the change in partial pressure necessary to effect this change is much smaller for CO_2. In the right graph, the concentration scale is expanded to make it possible to compare the media for which the capacitances are relatively small. Notice that $\beta_{w_{O_2}}$ is nearly 30 times smaller than $\beta_{water_{CO_2}}$ or β_{gas}. The lines in this figure are drawn from actual values at 37°C. Data on human blood CO_2 are from Christiansen *et al.* (1914) and data on O_2 are from Roughton (1964).

speed and direction of movement depends on the temperature and mass of the molecule and can be only be predicted statistically.

The difference in the quantity of gas that moves from one region to another by this random motion is the *net diffusion*. We are often interested in the quantity of gas that may diffuse across a given plane such as the surface of a gas exchanger so net diffusion is typically described in terms of the unit surface area across which the movement occurs.

One of the important factors that determines the rate of net diffusion must, therefore, be the area of the exchange surface. Temperature is also important, but because most biological exchange occurs without a temperature change, it is assumed constant for a given exchange.

The partial pressure difference across an exchanger describes the potential for net diffusion. When the partial pressures of a gas are the same on either side of an exchanger, there will still be random motion of molecules across it, but the quantities of gas moving will be equal in both directions and the net diffusion will be zero. The thickness of the barrier through which a gas molecule must travel impedes the rate of net diffusion. The thicker the surface of a gas exchanger, the more slowly net diffusion will occur. The ratio of partial pressure difference across a barrier to its thickness describes the diffusion gradient.

Each gas interacts with the medium through which diffusion is occurring in a characteristic way. This interaction is described by the diffusion constant that is specific to the gas and the medium. As an example, O_2 and CO_2 diffuse both absolutely and relative to each other at different rates in different media because each gas has its own physical characteristics and interacts in a specific way with the medium.

Combining the relationships just described, one can state that the rate of net diffusion ($\dot{M}$) is directly proportional to the diffusion constant (D), the exchange surface area (A), and the partial pressure difference (ΔP). It is inversely proportional to the thickness (t) of the exchange surface.

To keep the direction of the net diffusion mathematically consistent, the diffusion constant has a negative sign. In the discussion of diffusion, this mathematical nicety is often conveniently discarded. But, to write the equation for the rate of net diffusion correctly, it should appear as follows:

$$\dot{M} = -DA\frac{\Delta P}{t}$$

Respiratory gases are not actively transported across biological membranes. Gases move between media, across membranes, and short distances within media by net diffusion. Because of its passive nature, dependent only on the thermal energy of the gas molecules, diffusion is not sufficient for the transport of large quantities of materials over distances larger than a few millimeters. Movement of respiratory gases over larger distances requires the additional input of energy and bulk transport of the medium.

Bulk Flow

When a physical pressure is applied to a contained gas or liquid, it will have a potential to flow if possible, away from the source of the pressure in the direction of the least resistance. This statement describes the determinants of the form of mass transport called bulk flow. The pressures can be applied only to the entire medium, of which the gas is a part, not to the individual component gases. The flow rate (e.g., $\dot{V}$ or $\dot{Q}$) in liters/minute simply equals the difference in pressure (ΔP) divided by the resistance (R) so that

$$\dot{V} = \frac{\Delta P}{R}$$

In reality, bulk flows such as ventilation or perfusion of a gas exchanger can rarely be described so simply. The resistances and pressures are seldom constant. Nevertheless, the operating principle is sufficiently well founded to allow us usefully to understand, describe, and predict breathing or blood flow, for instance.

Putting the Three Fundamental Concepts Together

Capacitance describes the combination of a respiratory gas and its medium. Diffusion and bulk flow describe the means by which gases move from place to place. Although it is possible to derive many other relationships, none is a more fundamental determinant of gas exchange than these three.

These three concepts and some of their components share dimensions and are often interrelated, so it is not unexpected that incorrect applications occur. The confusion between the relevance of partial pressure and concentration to diffusion is a good example. Forster (1964) provided an elegant demonstration that should confirm your understanding of the three fundamental principles.

The experiment, shown in Fig. 1-4, examines diffusion of oxygen between two liquids separated by a membrane that is permeable only to O_2. The initial conditions have, on the right side, olive oil that contains dissolved oxygen at a concentration

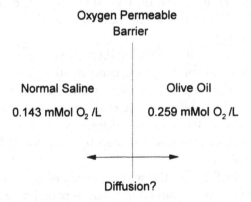

Fig. 1-4. Net diffusion will occur against a concentration gradient because it is dependent on partial pressure difference rather than concentration. In this case, the P_{O_2} in the saline solution is greater than that in the olive oil. See text for further elaboration. [Figure adapted and redrawn from Forster (1964)].

of 0.259 mM/L. An aqueous saline solution on the left contains oxygen at a concentration of 0.143 mM/L. With only this information, if asked to predict the direction of net diffusion, one might be tempted to say "From the olive oil to the saline, according to the concentration gradient."

But, having understood this chapter, one should say, "It is impossible to predict with the information given because there is no information about the difference in the partial pressure of oxygen across the barrier." Remember, net diffusion of oxygen is a function of ΔP_{O_2}. Concentration is not a component of the diffusion equation; it is, in fact, irrelevant to the prediction.

In order to predict the direction of net diffusion, it is necessary to know either the partial pressures or the capacitances that relate the concentration of the media to partial pressure. In the case of the two liquids in this example, the $\beta_{\text{olive oil}_{O_2}}$ is 48.9 μM/L kPa and the $\beta_{\text{saline}_{O_2}}$ is 10.7 μM/L kPa. The P_{O_2} in the olive oil is therefore 5.3 kPa, and in the saline it is 13.4 kPa. Net diffusion must and can occur only from the saline to the olive oil despite the concentration gradient in the opposite direction. At equilibrium in this example, the partial pressures of oxygen would be the same on both sides, but the concentration in the olive oil actually would have increased over that of the initial conditions.

In net diffusion between gases, because the capacitance is identical for all gases, concentration follows the same direct proportion to partial pressure in all cases. One can use concentration in the diffusion equation in this case but it is still partial pressure differences that determine it. It is best to use partial pressures in all cases to keep the conceptual basis for diffusion consistent.

In Fig. 1-5 the whole process of gas transport for both O_2 and CO_2 is diagramed in terms of partial pressure changes. This format is sometimes called a *cascade* because the partial pressures fall from the source of the gas. Note how the mode of transport differs in terms of the change in partial pressure. Where mixing of gases occur or diffusion occurs, there is a change in partial

O₂ and CO₂ Cascades

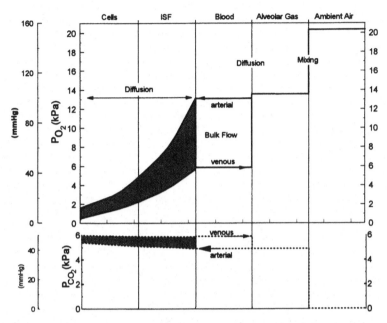

Fig. 1-5.   Oxygen and carbon dioxide cascades between the environment and cell. The partial pressure changes only where mixing or diffusional transport occurs. In the bulk flow phases of gas transport, there is no change in partial pressure.

pressure. When there is bulk flow, there is no change in partial pressure because the transport is a function of changes that involve the medium of transport rather than the abundance of the gases.

In the next chapter, the specific characteristics of the gases and media of respiration will be explained.

CHAPTER 2

PHYSICAL CHARACTERISTICS OF RESPIRATORY GASES AND MEDIA

This chapter contains many numerical values that should not be committed to memory. To write this section, I looked up all the numbers. The values are important and should be referred to when making calculations. It is more important to the understanding of gas exchange to keep in mind the relative magnitudes and their consequences. Everyone should be able to make rough estimates from memory. No one should rely on memory for important calculations. If you need the actual values, look them up, too. Otherwise, do not waste your brain storing facts; other media already do that so much better.

Diffusion

Physical constants that are relevant to diffusion for these two respiratory gases are given in Table 2-1. The mean speed of

17

Table 2-1. Physical Properties of O_2 and CO_2

	O_2	CO_2	CO_2/O_2
Mol. Mass	32	44	1.375
$1/\sqrt{mass}$	0.177	0.151	0.853
D (aka physical D) (cm^2/s)	0.178	0.138	0.775
β_{gas} (μM/L $\times$ kPa)	409.42	409.42	1.0
β_{water} (μM/L $\times$ kPa)	21.54	754.73	35.04
β_{water} D (aka Krogh's D) (nM/cm $\times$ kPa $\times$ s)	3.834	104.13	27.16

Note: All values at Σ0°C.

molecular movement of gases in a gas at a given temperature is inversely related to the square root of the molecular mass. Within that medium, the physical diffusion constant (D) for O_2 is larger than that for CO_2 by a proportion similar to that ratio: O_2 molecules diffuse more rapidly. Because the solubility remains the same throughout the exchange, diffusion is usually described in terms of concentration difference and the dimensions of D are cm^2/s.

When one considers diffusion between media, where the capacitance (β) values almost always differ, the apparent rate of diffusion will differ from that predicted from the value of the physical diffusion constant as defined above.

Between media, diffusion of gases can be predicted only from partial pressure differences. Krogh (1919) derived the diffusion constants used in gas exchange physiology to take this effect into account. His values are equal to the product of the physical diffusion constant and the solubility (capacitance) so that diffusion can be correctly predicted from the difference in partial pressure. After algebraic cancellation, the units of Krogh's diffusion constant are most conveniently expressed as nM/(cm $\times$ kPa $\times$ s).

In the case of gases moving between air and water, the relative speeds of diffusion are reversed. CO_2, because of its larger β_{water}, diffuses more rapidly than does O_2 and the value of Krogh's D is correspondingly larger.

The difference between the two forms of diffusion constant

is often confusing. That the diffusion of CO_2 can be described as both slower and faster than that of O_2 is not a contradiction. The two forms of the diffusion constant refer to different circumstances.

In air, where the β is the same for both gases, O_2 has a slightly larger diffusion constant than CO_2 because of the effect of molecular mass. In water, the value of β for CO_2 is so much larger than that for O_2 that the difference in the molecular masses has relatively little effect on the phenomenon.

Recall that β is the ratio of change in concentration to change in partial pressure. The effect of a larger value of β is that the partial pressure of gases changes less as a given number of molecules of the gas diffuse in or away. The effect on the diffusion equation is that the difference in partial pressures drops more slowly and the diffusion gradient is maintained. So, starting at equal partial pressure differences, for the diffusion of equal amounts of gas between water and air, for example, the partial pressure difference for CO_2 decays less than that of O_2. Using rounded values from Table 2-1 to calculate Krogh's diffusion ($D \times β$), one can see that the diffusive transfer of CO_2 between water and air should be faster than O_2 by a factor of roughly 30. For reference, values of β for O_2 and CO_2 in different media are given in Table 2-5 at the end of this chapter. Similarly, values for diffusion constants are given in Table 2-6.

Bulk Flow

In contrast to diffusion, changes in bulk flow depend solely on the characteristics of the medium of gas transport and are independent of those of the respiratory gas. Bulk flow is an solely an effect of physical pressure on the medium. Whatever affects bulk flow will affect the gas transport equally and without regard for the species of gas. If the flow of the medium past an exchanger is doubled then the amounts of CO_2 and O_2 carried are also doubled. Important differences between the various media that are relevant to bulk flow are given in Table 2-2.

Table 2-2.   Comparison of Gas Exchange Media

	Water	Air	Water/Air
Density (g/L)	999.841	1.293	773
Mass (g) of medium/mM O_2 carried	2309	0.146	15774

Note: All values at 0°C; P_{O_2} = 20.1 kPa.

From Table 2-2, the advantage of being an air breather, in terms of access to the external medium, can be seen. Water is almost 800 times more dense than air. The energy expenditure to move air via bulk flow is trivial relative to that required to move an equal volume of water. Beyond the difference in density, there is the 30-fold difference in β for oxygen, for example. The combined effects of density and capacitance mean that to gain access to a given amount of oxygen, the water breather must move almost 16,000 times the mass of the medium.

Composition of the Atmosphere

The continuous mixture of air masses and rapid diffusion of gas molecules generally prevent any changes in the fractional composition of the dry atmosphere, as given in Table 2-3, from place to place on the planet. Important changes may occur in confined environments, such as burrows, or where there are

Table 2-3.   Composition of the Atmosphere

Gas	% (dry)	% (wet)[a]
N_2[b]	79.02	74.13
O_2	20.94	19.65
CO_2	0.04	0.03
H_2O	0	6.19

[a] Saturated with water vapor at 37°C.
[b] Includes trace and other inert gases.

large influxes other gases, such as in some urban areas or near volcanoes.

Biological activity can affect the amount of nitrogen and oxygen in the atmosphere only slightly, and on a local and temporary basis. The mass of the atmosphere is simply too great. Until recently, the quantity of CO_2 that naturally entered the atmosphere approximately equaled the amount removed via sedimentation. Anthropogenic CO_2 is now estimated to be added to the atmosphere in amounts approximating that from natural sources (Budyko et al., 1987) so that the fractional amount of CO_2 is slowly increasing. Although this increase is not large enough to be directly important to individual gas exchange, it does have potential effects on a more global scale.

Water Vapor

Air always has some water vapor in it, even in the driest deserts. The presence of this gas dilutes the other gases and must be considered when working with the fractional composition of the atmosphere. Water vapor behaves rather differently from the other atmospheric gases because, unlike them, it exists near its triple point: at biologically tolerable temperatures, water may normally occur as solid, liquid, or gas.

The partial pressure and density of water vapor are indications of the *absolute* humidity of the air. At a given temperature, if the P_{H_2O} is raised sufficiently, the saturation of the air is exceeded and liquid water will condense. This saturation partial pressure increases nonlinearly with temperature as shown in Table 2-4 and shows that when the temperature of the air increases from 0 to 50°C, the amount of water vapor it can carry at saturation increases by a factor of about 20.

The *relative* humidity describes the percentage of the saturated vapor pressure that is measured in an air sample. Air that is near saturation at a cold temperature will have a much lower relative humidity as it is warmed, although its absolute water

Table 2-4.   Properties of Saturated Water Vapor in Air

Temperature (°C)	Vapor Pressure (kPa)	Density (g/m³)
0	0.61	4.9
5	0.87	6.8
10	1.23	9.4
15	1.70	12.8
20	2.33	17.3
25	3.16	23.0
30	4.23	30.4
35	5.60	39.6
37	6.25	43.9
40	7.35	51.1
45	9.55	65.6
50	12.30	83.2

content will be unchanged. This phenomenon accounts for the dryness in heated buildings as cold air enters during cold weather.

Nitrogen

The physical characteristics of air are fundamentally the same as those of its major constituent, nitrogen gas. One way to emphasize this point is to consider the mean molecular mass of air. From the molecular weight and fractions of the gases, one can calculate that value to be about 29, very close to the molecular weight of nitrogen.

We generally consider N_2 as though it were physiologically inert. In the chemical sense, that is generally safe. Before we are tempted to consign N_2 to physiological irrelevance we must note that the physical characteristics of air such as heat capacity, viscosity, and density are those of nitrogen. Replacement of N_2 in air with another chemically inert gas, He for example, has immediate and obvious effects on the respiratory system.

Oxygen

The earth's atmosphere is unique in the solar system because of the presence of free oxygen. Atmospheric oxygen has come almost entirely from the photosynthetic breakdown of water. Burning of fossil fuels has had a neglible effect on the concentration of O_2 in the air because of its large mass in the atmosphere. The density and other purely physical characteristics of O_2 are similar to nitrogen.

Carbon Dioxide

Like nitrogen, CO_2 in the atmosphere comes from outgassing of the crust. Unlike the other gases, CO_2 has a substantial reservoir in the oceans. The mass of CO_2 that exists in dissolved forms has been estimated to be about 50 times greater than in the air (Budyko et al. 1987). The cycling of CO_2 between the oceans and the air and through plants maintains the presence of CO_2.

Together with water vapor, CO_2 absorbs infrared radiation that would otherwise leave the earth. This effect is crucial to the maintenance of the temperature of the earth's surface and atmosphere within a range compatible with life. Although we are concerned about the potential warming of the planet from addition of CO_2 to the air, we must also recognize its importance in the prevention of cooling.

Table 2-5. Capacitance

| Temperature (°C) | Distilled Water | | | Air |
| | $\beta_{water_{O_2}}$ | $\beta_{water_{CO_2}}$ | $\beta_{water_{CO_2}}/\beta_{water_{O_2}}$ | β_{gas} |
	$\mu M/(L \times kPa)$			$\mu M/(L \times kPa)$
0	21.54	754.73	35.04	439.40
5	18.89	627.40	33.22	431.50
10	16.75	526.07	31.40	423.88
15	15.05	448.96	29.84	416.52
20	13.67	386.84	28.30	409.42
25	12.47	334.41	26.81	402.55
30	11.49	292.99	25.50	395.91
35	10.75	260.83	24.26	389.48
40	10.16	233.51	22.98	383.26
45	9.64	211.04	21.90	377.24
50	9.21	192.10	20.86	371.40

Table 2-6. Diffusion Constants

	D cm²/s	K nM/(cm × kPa × s)
O_2 through		
Air (0°C)	0.178 [a]	
Water (18°C)	19.8×10^{-6} [a]	2.70×10^{-7}
Muscle (20°C)	—	1.03×10^{-7} [b]
Connective tissue (20°C)	—	0.84×10^{-7} [b]
Chitin (20°C)	—	0.096×10^{-7} [b]
CO_2 through		
Air (0°C)	0.138 [a]	
Water (20°C)	17.7×10^{-6} [a]	
Water vapor through		
Air (0°C)	0.202 [a]	

Note: $K = \beta \times D$; for convenience of cancellation in this conversion, the volume dimension of β, usually expressed as a liter, is expressed as 1000 cm³.

[a] Data from Washburn (1929).
[b] Calculated from Krogh (1919).

CHAPTER 3

DESIGN OF GAS EXCHANGERS

The essential function of a gas exchanger is to bring together two different transport media. This juxtaposition must be such that exchange is rapid and extensive. At the same time, it must also keep the media separated to assure that they are not mutually contaminated and only the gases are exchanged. With these requirements in mind and the tools of the preceding two chapters at hand, the characteristics of the various kinds of gas exchangers animals use, and one they do not, can be analyzed with relative ease.

Separation of the transport media requires the critical step in gas exchange to be diffusion through some biological tissue. The factors that govern diffusion must then also describe the gas exchanger. Net diffusion of respiratory gases is, from the diffusion equation, directly proportional to the diffusion constant, the area of the exchanger, and the partial pressure difference across the exchanger. It is inversely proportional to the thickness of the exchanger. One may easily conclude that a good diffusional exchanger has the following characteristics:

1. It consists of material readily permeable to respiratory gases.
2. It has a large exchange surface area with simultaneous exposure to both media.

3. The partial pressure difference across the exchanger is maintained as large as possible.
4. The exchange surface is as thin as possible.

Diffusional Gas Exchangers

For microscopic organisms, the cell surface is a sufficient gas exchanger. Exchange with the unstirred interior of the organism is accomplished by diffusion alone so long as the diffusion path is about 0.25 mm or less (see Text Box 3-1). A larger "pure diffusion" organism could exist if the metabolically active part of the cell is arrayed at the surface as the case in some algal cells that contain large central vacuoles.

Larger simple organisms may ventilate internal cavities to keep inner cell layers in close contact with the environment. These latter examples point out that, for most multicellular organisms, there must be a bulk flow system to transport respiratory gases to the cells of tissues that are too thick for adequate gas exchange by diffusion alone.

Within most animals, the limitation of the diffusion distance from the gas exchanger's surface to the rest of the cells is overcome with circulation of body fluids. The limitation of the partial pressure difference at the environmental surface of the exchanger often requires a second bulk flow system (ventilation) to move the external medium. Thus, two different limitations of diffusion must be overcome with bulk flows to make the gas exchanger work.

In some cases, the solutions to these problems are more simple and elegant. Insects surpass both limitations with a single bulk flow system. Air is forced throughout the body and within diffusion distance of the cells via a system of tubes (tracheoles). Some salamanders that live in cold, well-aerated streams are lungless. The blood vessels in the skin provide a sufficient gas exchange surface while movement of the water maintains a high partial pressure difference. It also helps that their metabolism is low because of the low temperature. Most animals must rely on a double bulk flow system.

Text Box 3-1.   Limitations of Diffusional Gas Exchange

Dimensional limitations for gas exchange by simple diffusion, from Harvey (1928).*

1. Assume a spherical organism that uses O_2 at a rate such that P_{O_2} at the center = 0. (A $P_{O_2} \approx 1$ mmHg allows mitochondria to work so the model may not be unrealistic.)

2. Harvey rearranged the formula for diffusion through a "window" to that for the surface of a sphere and assumed that $C = P_{O_2} (0.21 \text{atm})$, D = Krogh's diffusion constant for tissue (1×10^5 cm^2 min^{-1} atm^{-1}), $\dot{V}_{O_2}$ = oxygen consumption (0.02 cm^{3}min^{-1}, a reasonable value for a protozoan), to solve for the maximum radius (r) of the organism as follows:

$$r = \sqrt{\frac{C(6D)}{\dot{V}_{O_2}}} = 0.25 \text{ mm}$$

3. Thus, all but unicellular organisms, some simple multicellular organisms, and embryos are excluded from sole reliance on diffusion for gas exchange from the environment to their innermost reaches.

---

* Lest we stand too much in awe of the mathematical and deductive skills of our predecessors, note that, in his paper, Harvey cites a colleague for doing the actual derivation for a sphere from work published by other authors. One of these, Fenn (1927), in turn, cites the aid of "an anonymous mathematician" for his derivation for a "long" cylinder (i.e., a nerve) that comes from the work of a previous researcher who originally did the derivation for a disk-shaped object. The trail leads back from there, but this example is sufficient to remind us that there are few wholly original ideas in science.

Aquatic Gas Exchangers (See Text Box 3-2)

Diffusive gas exchange occurs until the partial pressures on either side of the exchange barrier come to equilibrium. In the case in which both media have the same capacitance (β) and the same flow rate, the equilibrium partial pressure will be halfway between

---

Text Box 3-2.   Aquatic Gas Exchange in a Book about Mammals?

Note to the reader: Your interest may lie wholly in the gas exchange of an air breather such as the human, so why, you might rightly ask, should you to care about fish? I include this topic because it is a direct and simple way to emphasize that the nature of the gas exchanger is a function of the physics and chemistry of the gases and media and that the animal must conform to these rules to exist. So read about fish. You will learn something and you will understand the gas exchange of humans, or whatever creature you are interested in, more clearly and intuitively as a result.

---

them. The most effective gas exchange occurs when the equilibrium condition is not the mean of the two media but rather approaches that of the medium delivering the gas; that is, the medium removing the gas has received as much gas as possible from the medium delivering it. There are several possible ways to improve the gas exchange toward this ideal.

An increase in the flow rate of the delivering medium will help maintain a larger partial pressure gradient and thus improve diffusion across the exchanger. This mechanism is particularly effective when the delivering medium has a low β value, as is true for water carrying oxygen. Continued replenishment of the water flowing over, for example, a fish's gill surface, will help to keep the partial pressure gradient for O_2 high in a medium from which it can be readily depleted because of its relatively low solubility. A higher value of the β of the delivering medium will mean that more material can be transferred into it for a given change in partial pressure.

Beyond these changes in the media there is an anatomical adaptation that effectively improves exchange solely by its geometry. Because it is effective and adds no energetic cost to the organism, Schmidt-Nielsen (1972) has referred to it as "a cheap trick": the countercurrent exchanger. In this discussion, it will be useful to refer to Fig. 3-1 in which two models of exchange, concurrent and countercurrent, are illustrated.

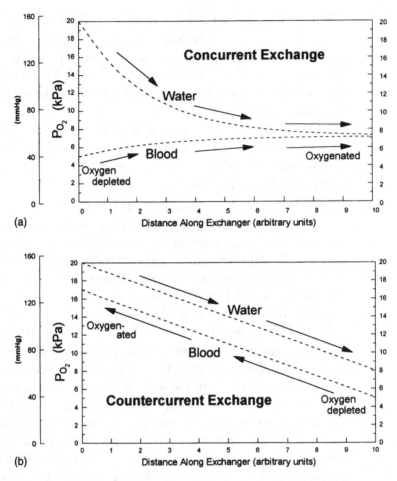

Figure 3-1.   Comparison of concurrent and countercurrent Exchange.

To add a more concrete context to the explanation, think of the illustration as a description of the exchange of O_2 from the water to the blood of a fish. The P_{O_2} of the water is 20 kPa in this example and the oxygen-depleted blood of the fish has a P_{O_2} of 5 kPa. Because of the partial pressure difference there will be some net diffusion as the water flows past the exchange surface in either case. If the flows of the two media are concurrent, the partial pressures will converge. As shown in Fig. 3-1a, the water, with its lower β, changes partial pressure more than the blood.

Although the gradient is large at first, the P_{O_2}'s converge at a value of about 7 kPa. The blood cannot gain appreciably more oxygen because there is almost no remaining partial pressure difference to yield net diffusion. If the flow of the water were increased in this example, the P_{O_2} of the blood could be increased, but there is an easier way.

If the direction of the blood flow is reversed, as is shown in Fig. 3-1b, the system becomes a countercurrent exchanger. The oxygen-depleted blood, with the lowest P_{O_2}, encounters the water at its lowest P_{O_2} but there is still a gradient for exchange. The water enters the exchanger at its highest P_{O_2}. The blood, which has picked up oxygen all along the exchanger, encounters the water that is just entering the exchanger and has the highest potential for diffusive transfer. In the countercurrent exchanger, there is always a good gradient for diffusion between the water and blood, so exchange continues as long as the flows are close together. In this illustration, the blood leaves the exchanger at a P_{O_2} of 17 kPa, more than double that of the concurrent model. Note that, just by changing the direction of the blood flow relative to the water, the blood leaves the exchanger at a seemingly para-doxically higher P_{O_2} than the water leaving the exchanger.

When water is the medium of exchange, the energetic cost of moving the mass of the medium and the advantage of the countercurrent arrangement appear to have required exchangers to be of a flow-through design; that is, the exit of the exchanger is separate and downstream from the entrance so that the momentum of the medium can be conserved as it is moved past the exchange surface. There are but a few examples in which the flow of the medium is stopped and reversed, as is typical in aerial gas exchange.

Aerial Gas Exchangers

The high β value of air for oxygen and the low energetic cost, relative to water, of moving the medium provided such an advantage that there was selection for the evolution of a new type of gas exchanger to allow air breathing. Although there are some fish that use their gills and gill cavities for aerial gas exchange, the delicate structure of the gill, which requires the buoyancy of the water for support, does not function well in air. The aerial gas exchange organ requires support to keep its delicate surface spread out in the presence of its less supportive medium.

The first vertebrate lungs appear to have arisen as outpocketings from the gut. They were kept inflated by the pressure of air that was, in essence, swallowed. This necessity forces a balloonlike design with a closure or valve to maintain the inflation. The mouth acts as a pump that inflates the lung. The elastic recoil of the lung and the weight of the other internal organs force it to deflate when it is opened to the outside. This system, still used in many lower vertebrates, directly employs the principles of bulk flow. When the air is sufficiently pressurized by the mouth, it flows into the lung. When the lung is opened, because the pressure inside is greater than that of the surrounding atmosphere, the gas inside flows out. Because of the mechanism for inflation, in which the pressure in the lung is always greater than that of the surrounding atmosphere, this system is referred to as *positive pressure* ventilation.

In higher vertebrates, bulk flow remains the mechanism for inflation of the lung but the means by which the pressure changes are achieved is different. The organs in the abdominal cavity can slide over each other because they are separated by a thin layer of liquid. When the volume of the cavity is expanded, the negative (relative to the atmosphere) pressure changes that result are transmitted through the fluid to all of internal organs and their contents, most of which are solid or liquid that is, incapable of expansion. Because the lung is the only organ with immediate access to air, this decrease in internal pressure causes air to flow into it from the atmosphere, inflating the lung. This *negative pressure* ventilation system has some distinct advantages.

No ventilatory muscles are attached to the lung itself because of the delicate nature of the exchange surface. The abdominal cavity can be expanded by muscular movement of the skeletal

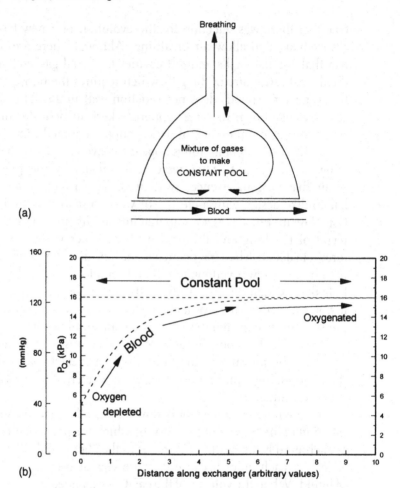

Fig. 3-2.   Model of constant pool type aerial gas exchanger.

elements, ribs and girdles, that surround it and by muscular displacement of the internal organs. The resulting negative pressure is transmitted to the lung uniformly and solely by relative negative pressure on the fluid that surrounds the lung wall. In a simple but clear sense, the lung is sucked open.

In most cases the lung never completely deflates. Its volume at exhalation is a function of the balance between its elastic recoil and a residual negative pressure that holds it open. The residual volume of gas that remains in the lung proves to be useful in this type of gas exchanger.

The incomplete deflation of the lung is a feature rather than a flaw in the design. It leaves a residual volume of gas that is mixed with the fresh gas brought into the lung. Turbulent flow and rapid diffusion of gas molecules generally assure thorough mixing. If the residual volume is large relative to the volume of the breath, the composition of the gas in the lung changes little during ventilation. This type of ventilation results in what is called a *uniform* or *constant pool* gas exchanger. The high β of gas for oxygen and frequent refreshing with outside air mean that, unlike the concurrent aquatic exchanger, the P_{O_2} of the delivering medium remains high throughout gas exchange. The blood equilibrates with that high P_{O_2} as is illustrated in Fig. 3-2.

In the constant pool model, the direction of the blood flow is irrelevant. It is more important that there be a good exposure of the blood to the gas. In most cases, unlike the simple model in Fig. 3-2a, the blood flow surrounds the chamber that contains the gas.

Nearly complete equilibrium is achieved between the media of the constant pool gas exchanger because the gas and blood involved both have high β values and the time of exposure is adequate. Nevertheless, the P_{O_2} of the blood leaving the exchanger can never exceed that of the gas in it as is the case in the countercurrent exchanger. The analysis in Text Box 3-3 presents another way to compare the efficacy of gas exchangers.

Text Box 3-3. Efficacy Ratio

A mathematical representation of the efficacy of gas exchange models was developed by Piiper and Scheid (1972). Further developments of these analyses will be considered in a later chapter but their original derivation will suffice for this introduction. The efficacy ratio is defined as a ratio of the partial pressures of gases entering and leaving the exchanger as follows:

$$\text{Efficacy} = \frac{P_e - P_a}{P_i - P_v}$$

In this equation, the subscripts refer to oxygenated or arterial (a) and oxygen depleted or venous (v) blood and to inhaled (i) or exhaled (e) external medium. The highest efficacy would result when the ratio equals -1. In that case, the arterial and inhaled P_{O_2} would be equal because of complete exchange, and a similar equality would occur between the venous and exhaled P_{O_2}.

The concurrent and constant pool exchangers can have a theoretical maximum ratio of zero when there is no diffusion limitation and flows of the medium are optimum. In that unlikely case the arterial P_{O_2} and exhaled P_{O_2} would be equal. In practice these exchangers will always have $P_a < P_e$ so the value of the ratio will always be greater than zero. Of the models discussed in this chapter, only the countercurrent exchanger can achieve a $P_a > P_e$ that would give a negative number. Taking the values from the examples in Figs. 3-1 and 3-2, the efficacy ratios are, respectively, concurrent, 0.013; countercurrent, -0.6; constant pool, 0.003. The constant pool is more effective than the concurrent exchanger, but neither is as effective as the countercurrent exchanger.

CHAPTER 4

FUNCTIONAL ANATOMY AND VENTILATION OF THE MAMMALIAN LUNG

The ventilation of a mammal's constant pool exchanger poses some interesting challenges for our understanding. From the perspective of the simpler systems of lower vertebrates, it may seem cumbersome and needlessly complex. Consider the following primary problems as you read through this chapter. Reciprocating flow necessitates "contamination" of inhaled air with gases already exposed to the exchange surface. Portions of the system (somewhat morbidly referred to as *dead space*) that do not function in gas exchange must be ventilated before the exchange surfaces can be reached by outside air. The thin exchange surface is not supported physically by the exchange medium, as is the case for water breathers or air breathers with positive pressure ventilation. The inhaled gases must be warmed and humidified to protect the delicate exchange surface. The air passages and circulatory vessels must be completely intertwined and the flow to them must be matched to create effective gas exchange.

As further obstacles to our understanding, the fluid mechanics of air flow is not a simple subject and may not be intuitively obvious. Finally, an often-neglected consideration of part of a living system is that the lungs also have nonrespiratory functions and may vary in both structure and function from one individual to the next.

Conducting Airways

The structure of the upper portion of the ventilatory system is diagramed in Fig. 4-1. To add function immediately to this anatomy, consider that the major role of these portions is to condition the gases and conduct them to the regions of gas exchange. Except for the addition of water vapor to the inspired air, no important gas exchange occurs in the conducting airways. Their relatively thick walls, small surface area, and poor blood supply combine to make these passages ill-suited for diffusive exchange despite the large airflow through them.

The nostrils and nasal hairs perform an initial coarse filtration of the inspired air. The airstream next encounters the walls of the upper airways that are maintained as moist surfaces by mucous secretions. The turbinate bones support additional moist surfaces that project into the airways. As the word *turbinate* suggests, the airflow may be tortuous and turbulent through these passages and into the pharynx. In this swirling flow the momentum of larger (diameter > 10 μm) particles of suspended matter causes them to collide with the walls where they adhere to the mucus. Cilia propel the mucus to the back of the pharynx where it is swallowed or expectorated (Newhouse et al., 1976). At the same time as this filtration is happening, the air is warmed and humidified, that is, *conditioned*, appropriately for the exchange surfaces. Breathing through the mouth results in lower resistance to flow (see Text Box 4.1) at the cost of less effective air conditioning.

Below the glottis, the conducting portions of the airways

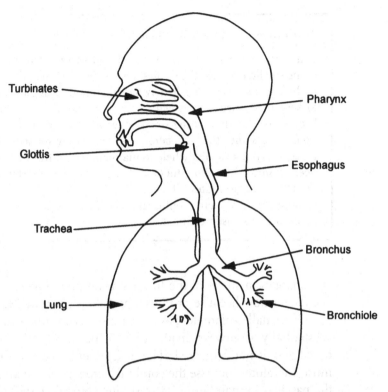

Fig. 4-1. The nasal passages, mouth, pharynx, trachea, bronchi and upper portion of the respiratory tree. This sketch, from a frontal view of a dissected human lung, is intended to emphasize, in two dimensions, the irregular and asymmetrical branching of the airways, rather than accurately portray the anatomy. Note, however, there are three branches for the right bronchus and two for the left. The location of the heart and aorta result in the two lungs' asymmetry. The right lung has three lobes, the left two. The lobular nature of the lung contributes to its flexibility and differential ventilation of regions during breathing. Only the first few of many generations of the conducting airways are shown.

include the trachea, bronchi and bronchioles that divide repeatedly to produce the treelike pattern that will reach regions where gas exchange can occur. These passages are approximately circular in cross section and are supported with cartilaginous rings that allow the tubes to flex without kinking. Movements of the neck

---

Text Box 4.1.   Airway Resistance

Unlike the circulatory system, where the greatest resistance occurs in the smallest vessels, the greatest resistance to ventilation occurs in the larger airways. People may flare their nostrils as they inhale to keep the passageway as wide as possible. Some athletes have been seen recently wearing small adhesive bandages across their noses. For some, this affectation may serve only for style, but, for others, it may actually help to maintain the nostrils in a more open position and allow them to breath with less resistance at normal ventilation rates. Of course, with any exertion, they will breathe through their mouths and the bandage is once again reduced to a stylistic element.

---

and changes in shape of the lung as it fills and empties will not occlude airflow because of the support the rings provide.

For smaller-diameter passages, resistance to airflow increases substantially as airways divide, even if the combined cross-sectional area is not decreased. The division of airways in the lung must, therefore, increase the combined cross-sectional area of all the passages to maintain a low-resistance bulk flow system. The practical consequence of this principle of fluid dynamics is clearly illustrated in the dimensions of the airways as division precedes.

The diameter of the human trachea is about 16–20 mm so its cross-sectional area is roughly 200–300 mm^2. Beyond the bronchi, the branching of the respiratory tree is not entirely regular or symmetrical, so its idealization as generations of bifurcations may not be an appropriate model. Nevertheless, the divisions of the conduction airways are described as continuing through about 15 or 16 generations. Careful measurements of lung casts (Weibel, 1963; Horsfield and Cumming, 1968) suggest that the combined cross-sectional area of the smallest conducting passages, the terminal bronchioles with a diameter of about 0.7 mm, is on the order of 100 times greater than the trachea.

The flow rate of air through a tube may also be analyzed as the product of velocity and cross-sectional area. As the inspired air is distributed through the increased area, the velocity must decrease. This decrease in velocity has three important consequences. First, the slower flow is less likely to be turbulent. Second, at very slow bulk flows, diffusion becomes more important in the ultimate transport and mixture of respiratory gases. Third, particulate matter that had escaped filtration will settle on the walls because the ability of a fluid to carry matter in suspension decreases with decreased velocity.

Exchange Airways

As one should rightly expect, the qualities of the exchange portion of the airways, the respiratory bronchioles, alveolar ducts, and the alveoli are those of a good model gas exchange surface. These structures have thin walls, a large surface area, and a close proximity to the other exchange medium, the blood.

The exchange portions of the airways that come from a given terminal bronchiole are referred to collectively as an *acinus*. Although it is common to represent the arrangement of the alveoli as similar to grapes in a bunch, the structure is better suggested by a sponge or honeycomb. The alveoli do not exist as little balloons on stems but as a set of adjacent polyhedral compartments that open to a common airway. The structure of the acinus is diagramed in Figs. 4-2 and 4-3.

Adjacent acini may be ventilated via entirely separate branches of the airways. The movement of gases may occur via alternate routes that can compensate for obstruction of the more direct ventilatory pathways. Interalveolar pores allow communication between alveoli and anastomoses of airways may allow collateral ventilation between adjacent segments of the lung even though the bronchi from which they originate are blocked (Henderson et al., 1969). The interchange of gases between acini also helps to ensure the uniformity of the composition of the gases in the lung.

The dimensions of the acinus are important to understanding the gas exchange in the lung. The alveoli are of barely visible

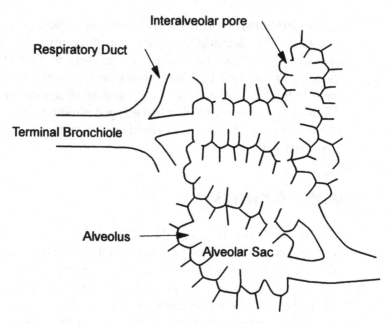

Fig 4-2. Portions of two adjacent acini. The diagram is intended to emphasize that the tissue resembles a diffuse sponge more than a set of balloons. Notice that there are openings between the acini that would allow collateral ventilation if the air way were blocked.

size, about 100–300 μm across, and the communicating airways are roughly double that. It is in dimensions like these that diffusion can again become an important mechanism for gas transport. Diffusion accounts for the rapid uniformity of gas composition that occurs within the acinus (Cumming et al., 1966). The gases that are introduced and removed via the blood and ventilation are rapidly distributed throughout the volume. Although slow, bulk flow does not disappear at these levels because, after all, the volume change of the acinus during breathing cannot come from diffusion.

Volumes of the Lung

The total volume of the lung (TLV) at maximum inspiration is a function of body volume. For convenience, we use a representa-

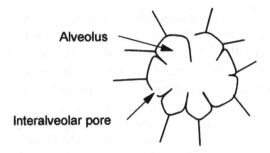

Alveolus

Interalveolar pore

Fig. 4-3.   Diagrammatic cross section through an acinus.

tive value, say, 6.0 L, but it should be understood that the individual variation is great. During normal ventilation, the volume of the lungs is much less than their maximum; they are slightly less than half-inflated. The amount of gas that moves in an out of the lungs during breathing at rest, called the *tidal volume* (V_T), is smaller still, about 0.5 L.

The volume change of the lungs during inspiration of the tidal volume occurs primarily in the exchange portion of the airways. Therefore, the 0.5 L of gas that flows through the conducting airways consists first of that gas that remained in the conducting airways from the previous expiration and then of air from outside the body. The air reaching the exchange portion of the lung will, therefore, be lower in oxygen and higher in carbon dioxide and water vapor than the outside air. As a breath is exhaled, the first gas to leave the body is that from the conduction portions of the system. Except that it is probably warmer and

---

Text Box 4-2.   Lung Volume and Body Size

Average lung volumes are usually directly related to lean body mass of the individual. With increasing age (after attainment of adult body size), total lung capacity remains relatively unchanged but residual volume increases by about half from age 20 to 60 at the expense a decreased vital capacity. Adult females average about 10% less total lung volume relative to males of the same height. In both genders, variations of as much as ±20% of the average are considered normal.

more humid, it does not differ from the outside air. The last air of the breath shares its composition with that of gases from deep within the lung, the region where gas exchange has occurred.

The tidal volume is conventionally described as two discrete parts although there is some mixing between them. The volume of air from the conducting airway is called the *dead space volume* (V_{DS}) and that from the exchange portion of the airways, the *alveolar volume* (V_A). V_{DS}, about which much more will be said later, is about 30% of a resting tidal breath, that is, about 150 ml in our example.

The volumes of gas that remain or "reside" in the lung while we breathe are the *residual volumes*. After a maximum exhalation, the lung is not emptied of gas. There remains, because the ribs prevent the lungs from complete collapse, a *residual volume* (RV) of a little more than 1 L in our example. This volume, combined with the gas that is not exhaled after a normal tidal breath, makes up the *functional residual capacity* (FRC), that volume that dilutes the incoming breath of air.

The FRC is usually about four to five times larger than the tidal volume so its composition is not much affected by the tidal dilution. This constancy of the composition of the alveolar air results in similar constancy of the partial pressures of respiratory gases in the arterial blood that is in near-diffusional equilibrium with it. Despite the seeming disadvantage of "contamination" of the fresh air when it is mixed with the FRC, the constant pool gas exchanger can now be seen to provide the advantage of an effectively constant environment for gas exchange within the body.

There are other measures of lung volume that will be important to understanding its function. The volume change between the deepest inhalation and the deepest exhalation is the *vital capacity* (VC). Obviously then, the total lung volume can be described as the sum of the vital capacity and the residual volume (TLV = VC + RV). The amount of gas that can be exhaled after a normal breath is the *expiratory reserve volume* (ERV). Algebraically, ERV = FRC − RV. Similarly, the *inspiratory reserve volume* (IRV) is the volume that can be inhaled after a

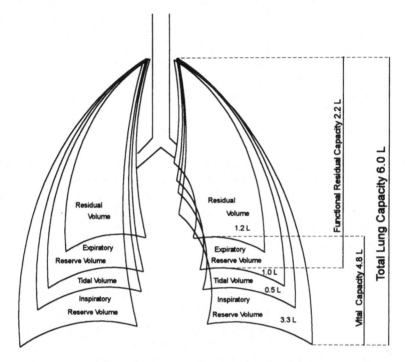

Fig. 4-4.   Lung volumes and capacities.

tidal inhalation. Algebraically, it is more tedious than I care to describe here, but I assume my readers can figure it out if they so desire. Fig. 4-4 illustrates the relationships of some of the lung volumes in graphic form that may help clear any confusion from the verbal description.

CHAPTER 5

STATIC AND DYNAMIC MECHANICS OF THE LUNG AND CHEST

The lungs are not a simple tissue. Explanations of the mechanics of breathing have been difficult because of the difficulty in relating the properties of the lung to familiar common materials and because, in many cases, the structure of the lung has been incorrectly understood. Lungs are not perfectly elastic and their properties are poorly represented by the models that have been used. Most models rely on positive pressure for inflation and treat the alveoli as independent, bubblelike devices. In reality, neither of these assumptions is correct. Both lead to the misunderstanding of how the lungs work.

The lungs are inflated by negative pressure and do not resemble balloons at all in their structure. The responses of the lungs to the forces of ventilation are a complex interplay among the mechanical properties of their spongy or frothlike tissue (*parenchyma*) and the effects of surface tension of the fluid lining of the airways. Attempts to resolve these properties in terms of simple mechanical models like balloons and soap bubbles have led to treatments of the subject that may be more readily understandable but bear little resemblance to the actual structure and function of the tissue.

Mechanical Properties of the Lung and Chest Wall

When the exchange airways expand in volume as the lung fills, they do so because the surrounding tissue uniformly pulls the walls of the alveoli and exchange airways outward. The expansion creates a relative negative pressure that causes outside air to flow in. The force on the walls, called *radial traction*, results from the elastic properties of the walls and the fluids lining them. Radial traction pulls from all directions on all the airways and blood vessels within the lung. Airways that are less elastic such as the cartilage-reinforced conducting airways expand little, if at all. Thin-walled exchange surfaces of the airways and capillaries are more elastic and stretch open in response to radial traction. This phenomenon is illustrated in Fig. 5-1.

From the paragraph above, it should be apparent that, in their normal condition, the lungs are held stretched open by an external force. The forces of radial traction that are distributed throughout the parenchyma of the lung are maintained by the negative (relative to atmospheric) pressure between the surface of the lung and the chest wall and diaphragm that keeps the lung stretched to fill the thoracic cavity (see Fig. 5-2). The pleural membranes that line the lung's outer surface and the inner wall of the chest slide over each other during breathing. The sliding is made easier by the presence of a small volume of intrapleural fluid, occupying the so-called *intrapleural space*, between them. It is this fluid that actually transmits the negative pressure between thoracic walls and the surface of the lung because the pleura connect to the surrounding tissue only where the bronchi enter

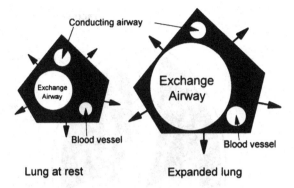

Fig. 5-1. Radial traction. Tension from all sides acts on the parenchyma so that each tissue expands in accord with its elasticity. In this two-dimensional representation, the exchange airways are the most elastic and the conducting airways the least.

the lung. In intuitive, if not physically rigorous, terms, the chest wall simply *sucks* the lung open. Intrapleural fluid slowly secreted into the intrapleural cavity is removed by lymphatic pumping, so the relative negative pressure of the intrapleural space is maintained.

The space occupied by the gases in the lung may be called *pulmonary, intrapulmonary,* or *intra-alveolar.* For simplicity I will use the term, pulmonary, to indicate this space. Even at rest the pressure in the intrapleural space must be less than on the pulmonary gases. The elastic forces the lung tissue generates as it is stretched open would cause the lung to fall away from the chest wall otherwise. This idea leads to a simple rule: The intrapleural pressure, regardless of the action of the lung, must always be lower than the pulmonary pressure. If it were not so, the lung would simply begin to deflate because nothing would oppose the elastic forces.

Any connection to the outside atmosphere and the intrapleural space will break the liquid seal and result in a *pneumothorax.* The lung becomes free to recoil and begins to collapse as air fills the expanding intrapleural space (see Fig. 5-3). The connection need not necessarily be through the chest wall. It can also result from an internal rupture through the pleural membrane into the

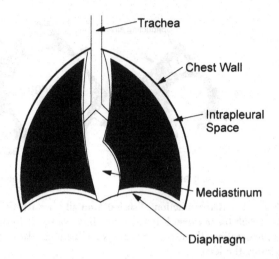

Fig. 5-2. **The lung and chest wall. The size of the intrapleural space is exaggerated in this figure.**

parenchyma. Bleeding into the intrapleural space, *hemothorax*, will have the same result: The lung will tend to collapse as the liquid flows in and allows the intrapleural space to expand.

Fortunately, in humans, the *mediastinum*, containing the heart, large blood vessels and other thoracic organs, completely separates the right and left lungs so that if one lung collapses, ventilation of the other can continue (Fig. 5-4). In animals that have only partial mediastinal separation (dogs, for example) any pneumothorax will cause both lungs to collapse. When a human's chest cavity is opened as in heart surgery, the lungs must be kept inflated and ventilated by externally applied positive pressure.

After a pneumothorax, the lung is said to collapse. The inference of this statement is not entirely true. Although the lung rapidly loses volume, it is not reduced to a mass of tissue completely devoid of air. The small airways tend to collapse before the alveolar sacs are completely empty and air is trapped in the acini. The entrapped air prevents the lung, once filled, from ever completely emptying of air. The airways must be connected to a vacuum to remove all the air from them.

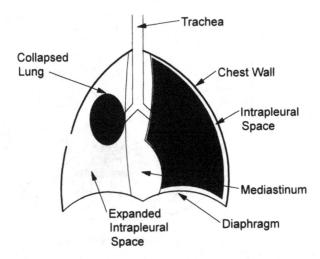

Fig. 5-3. Pneumothorax. Opening the intrapleural space allows the lung to collapse until the pressure of the air trapped inside equals the elastic forces of recoil.

Pressure, Volume, and Flow

The resistance to flow in the airways is very small during resting ventilation. This can be attested to by the small changes, usually ± fractions of 1 kPa, in pulmonary pressure that are observed as the tidal volume moves in and out (Fig. 5-5). The overall resistance to inflation of the lung comes only in part from the resistance to flowing air. The expansion of the lung is resisted by the elasticity of the parenchyma and the surface tension of the liquid lining the exchange surfaces. Further, the ventilatory muscles must also work against the mass and elasticity of the chest wall.

Static Pressure Volume Relationships

The classic work of Rahn et al. (1946) examined the separate and combined aspects of the lung and chest wall pressures at static

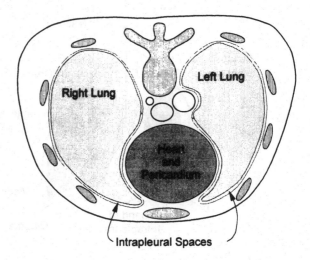

Fig. 5-4. Cross section of the thorax to illustrate the relationship of the pleural membranes and the mediastinum to the lungs.

volumes. At rest, assuming the airways are open to the atmosphere, the pulmonary pressure will be equal to the atmosphere, that is, zero. The resting volume of the lung occurs at the position at which the tendency of the lung to recoil just balances that of the chest wall to expand. After a pneumothorax then, the chest cavity will expand slightly as it is released from the inward pressure of the lung.

The resting volume of the chest cavity alone, without the inward pressure of the lungs to reduce it, will be larger than the volume of the intact system. In the case of diseases like emphysema that weaken the parenchyma, the resting volume will tend toward that of the isolated chest wall, approaching 75% of the vital capacity. The resting volume of the lung isolated from the chest, as we already know, will be much smaller than the combined resting volume. These relationships are illustrated in Fig. 5-5.

A pneumothorax that affects one lung must increase the residual volume of the remaining lung at the expense of its inspiratory reserve capacity. The expansion of the chest is opposed by the elastic recoil of only one lung in this case, so the rest position

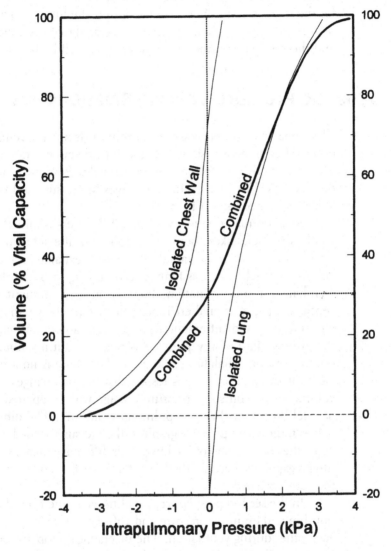

Fig. 5-5.   Contributions of the lung and chest wall to the combined relaxation pressure. Curves constructed and smoothed from the data of Rahn *et al.*, 1946.

of the system will be at a larger volume than would be the case for the intact system. The ability to empty the lung, also a result of the combined mechanics, will be decreased for the same reason and the residual volume of the intact lung will be larger.

Dynamic Pressure Volume Relationships

The actual changes in pressure and volume during the ventilatory cycle differ from the static relationships described above. It is often instructive to consider these variables from two different perspectives. First, we examine changes in pressure and volume as a function of time (Fig. 5-6).

At a normal frequency of about 12 breaths/min a single ventilatory cycle takes about 5 s. Inspiration, the active phase of ventilation occurs more rapidly the passive exhalation and peak volume is reached in less than 2 seconds (Fig. 5-6). At the end of inhalation and exhalation, there is, at least momentarily, a ventilatory pause when there is no airflow. The changes in pulmonary pressure and intrapleural pressure are appropriate to drive the airflow. Pulmonary pressure is negative during inhalation, positive during exhalation, and zero when there is no airflow.

Intrapleural pressure is responsible for the changes in lung volume so its course of pressure change will be approximately parallel but lower than the pulmonary pressure. The difference between these two pressure graphs is the pressure needed to overcome the elastic recoil of the lung. This difference increases as the lung is expanded because the lungs' tendency to recoil increases as they are stretched.

In the second perspective, we eliminate time from the relationship and examine pressure and volume as they relate only to each other during ventilation. The mechanical property, *compliance*, describes the change in the lungs' volume relative to change in pressure. It is not only nonlinear but differs depending on whether the lung is inflating or deflating. The dependence of the dynamic compliance of the lung on the immediately preceding events in the cycle gives rise to a *hysteresis curve* as shown in

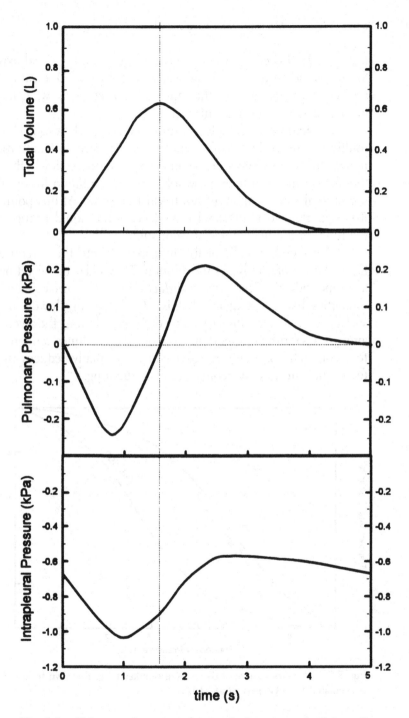

Fig. 5-6. Volume and pressure changes during a normal tidal ventilatory cycle.

Fig. 5-7. In this curve, there are two values on the vertical axis for any value on the horizontal axis and vice versa. One cannot predict one variable from the other without knowing what the immediately preceding events were.

The hysteresis of the lung pressure-volume relationship has multiple causes. The tissue, once it is stretched (inhalation), returns to its unstretched (exhalation) position more slowly than the change in pressure alone would cause. The liquid lining of the lung also has important nonlinear properties. At this point, it is important to understand the effects of the liquid lining of the lung on the mechanics of ventilation

The secretory cells of the lung release into the alveoli a liquid that is not simply a saline solution. They add to the secretion a compound called *pulmonary surfactant* that has surface active properties like a detergent, that is, it substantially lowers the surface tension. The surface tension of the liquid lining the alveoli and airways forms an important component of the lungs' tendency to recoil. The presence of surfactant substantially reduces the forces that must be overcome to inflate the lung.

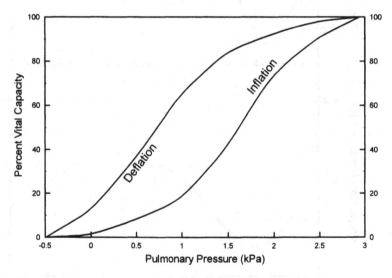

Fig. 5-7.   Hysteresis curve of the excised monkey lung. Redrawn from data recalculated from Glaister *et al.* (1973).

In the airways and in the junctions formed where the alveolar surfaces meet, the relationship between the curvature of the fluid surface and the force generated by surface tension can be described by Laplace's law. Unfortunately, Laplace's law has been applied to the individual alveoli of the lung in a misleading way. Applied to the pressure inside a bubble, Laplace's law correctly indicates that the pressure inside the bubble is inversely proportional to the radius: Smaller bubbles have higher pressure than larger bubbles. If it is assumed that the alveoli are correctly modeled as independent little wet balloons attached to hollow stems, then Laplace's law can be invoked to suggest the smaller alveoli should, because of their inherently higher pressure, collapse and inflate the larger alveoli. That this collapse does not occur is credited to the presence of the pulmonary surfactant that, in all cases, perfectly alters the surface tension as a function of the radius of each alveolus so that Laplace's law is thwarted.

But, the alveoli are not bubbles supported by pressure from within. Neither are they spherical or independent of each other. In the parenchymal "froth," the walls of a given alveolus are structurally parts of several adjacent alveoli. The connective tissue within the alveolar walls causes each alveolus to be supported by its transmural neighbors. As one alveolus changes its volume, it does so in interaction with the stresses on the adjacent alveoli. The reduction of surface tension by the surfactant is extremely important to the reduction of the pressure necessary to inflate the whole lung as is illustrated by the difference in the curves for air-filled and saline-filled lungs (Fig. 5-8). The detergent effects of surfactant are important to the patency of the small airways but exert a rather small force on the curvatures within the alveoli. It is not clear that the surfactant is in any way important in the maintenance of the patency of individual alveoli.

The maximum pressure that the lung can exert when exhaling against a closed glottis is usually no more than about 20 kPa. Maximum static inspiratory pressure is around -10 kPa. There is a dramatic consequence of this pressure; see Text Box 5-1. During dynamic breathing, the pressures do not reach these extremes. The pressures exerted are related to the volume in the

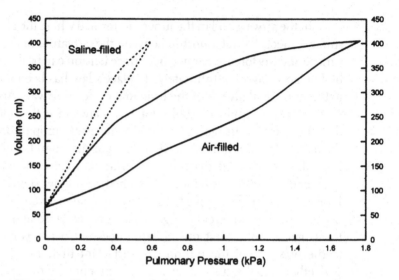

Fig. 5-8. Hysteresis curves of air- and saline-filled excised monkey lung. Redrawn from data recalculated from Bachofen *et al.* (1970).

Text Box 5-1.   Inspiratory Pressures

In some early adventure films the hero was supposed to have escaped his pursuers by hiding out of sight beneath the surface of a convenient body of water while breathing through a hollow reed. The pressure exerted by the water increases by about an atmosphere, 100 kPa, for every 10 m of depth. If the maximum inspiratory pressure is about −10 kPa then the maximum depth at which one could still inhale is theoretically about 1 m. As those who use a snorkel for skin diving know, the maximum practical depth is much less than a meter. As a result snorkels are short and wide enough to pose little resistance to ventilation. Our hero trying to breath through a long narrow reed would have found out quickly that he would have been better off to learn some respiratory physiology and try to hide elsewhere.

The same principle means that long-necked dinosaurs could not have walked the bottoms of lakes while bringing their heads to the surface to breathe. They could not have inhaled against water pressure on the outside of their bodies at that depth.

lung because of the combination of muscular and elastic forces that varies with inflation. To examine the dynamic relations of flow rate and volume in a normal lung, several indices are used. These measures involve either flow rate as a function of time or of the lung volume.

If one inhales to vital capacity and exhales as forcefully as possible while recording the expired volume as a function of time, a characteristic curve is produced (Fig. 5-9). An individual with a normal lung will exhale about 80% of the forced vital capacity (FVC) in 1 s. This measure is referred to as the forced expiratory volume ($FEV_{1.0}$). If the airways are obstructed, the rate of flow will be slower and so the $FEV_{1.0}$ will be less than normal, reduced to perhaps 40% of vital capacity. If the movement of the lung is restricted by conditions that make it or the chest less compliant, the $FEV_{1.0}$ may be even greater than normal, in terms of fraction of vital capacity but the vital capacity itself will be less than normal because of the restriction.

The case of restriction emphasizes that it is not just the muscular effort that empties the lung; the elastic recoil also contributes to the pressure of emptying. If the relationship of flow rate

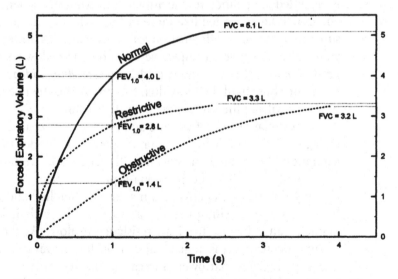

Fig. 5-9.   Forced expiratory volumes ($FEV_{1.0}$) for normal and impaired lungs.

to volume is examined, this principle is made more obvious. Again, measuring a forced expiration from vital capacity, we employ an instrument that plots flow rate against volume to get characteristic curves for normal and impaired lungs (Fig. 5-9). Note that the maximum flow rate is achieved at maximum volume and that, as the lung is emptied, the flow rate decreases, reaching nearly zero as the residual volume is approached. This decrease in flow occurs despite the continued effort of the individual to exhale as forcefully as possible. At the smaller volumes not only is the recoil of the lung less but also the muscles must attempt to compress the chest wall while its resistance to compression increases.

An individual with obstructed airways will, of course, not be able to match the flow rates of the unobstructed case. Chronic obstruction has further effects that will be discussed below. The restricted individual will display data that confirm the limited vital capacity with perhaps a greater flow rate than normal when the curves are matched to the same volumes.

The flow volume curve can be useful to understand another principle of lung function. If a series of maximal effort expirations is recorded that differ in the initial lung volume, a family of curves will be generated like those in Fig. 5-10. If the curves are superimposed, the different curves all coincide as lower lung volumes are reached irrespective of effort. The down sloping portion of the curve can, thus, be described as *effort independent*. No matter how hard the individual tries to exhale, the maximum flow rate will be the same at a given lung volume.

This relationship of volume and flow occurs because of the collapse of airways at high expiratory pressures. Under forceful expiration, the intrapleural pressure will become positive although still usually less than the pulmonary pressure. Starting at vital capacity the lung parenchyma is maximally stretched and radial traction is at a maximum so the airways will have their largest diameter and, hence, the least resistance to flow. As the lung volume decreases, the radial traction on the airways decreases and they become narrower, increasing the resistance to flow. Greater effort to expire only increases the pressure on them from

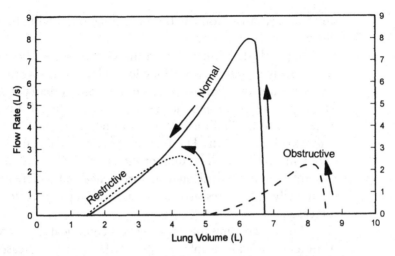

Fig. 5-10.   Forced expiratory flow rate as a function of lung volume.

the surrounding tissue and causes the airways to become still narrower so no greater flow results. For this reason the effort independent part of the flow-volume curve is the same for at a given volume regardless of where the expiration began.

Work of Breathing

Physical work, which we normally think of as the product of force and distance, may also be expressed as the product of pressure and flow rate. If we measure the pressure and flow rate during various levels of ventilation, we can calculate the fluid mechanical work of breathing.

In addition to overcoming resistance to airflow, the ventilatory muscles must also expend energy to overcome the elastic recoil of the lung and to move the mass of the tissues involved. These latter contributors do not increase as dramatically as fluid resistance with increased flow rates. It is possible to subdivide the total work into components that reflect the various mechanical contributors to ventilatory work. At the risk of the inherent confusion that can arise from such an examination, I will attempt to illustrate these components.

Fig. 5-11 has two scales on the vertical axes. The left axis qualitatively represents work done for each breath. The right axis represents the total work done per minute, rather than per breath. The right axis is the product or the work per breath times the frequency. With that distinction in mind, consider first the resistive work per breath. The increased frequency means the flow rates will be higher. The drag of the air in the airways and the amount of energy dissipated in turbulence both increase with air velocity, so the resistive work increases rapidly as frequency increases.

The explanation for elastic work is complicated, but for this treatment, we can simply say that as frequency increases, the

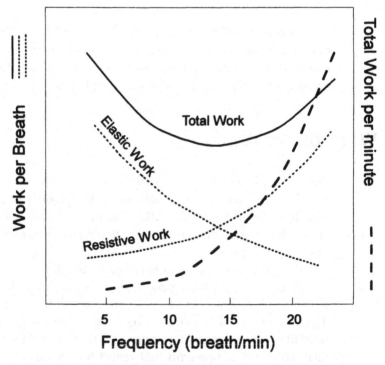

Fig. 5-11.   Work of breathing as a function of ventilatory frequency.

recoil of the stretched tissues returns more elastic energy to the process of movement. Hence, the elastic work per breath decreases with increased frequency. The total work per breath is the sum of the resistive and elastic work. As is obvious from Fig 5-11, total work is least at intermediate frequencies (approximately at the resting ventilation frequency) and higher at high and low frequencies.

The total work per minute, plotted as a function of the right axis of the figure, increases at higher frequencies. Multiplication of the work per breath by the increasing frequency results in a curve that increases rapidly at high frequencies. The work rate of breathing, trivial at low rates of ventilation, can thus become a substantial fraction of total metabolic rate at high levels of exertion.

CHAPTER 6

Ventilation and Perfusion of the Lung

The lung is perfused by the right side of the heart. The pulmonary capillaries offer less resistance than the systemic capillaries, so the systolic and diastolic pressures in the pulmonary artery are much lower than in the aorta. This simple consequence of the principle of bulk flow has some important effects that are especially manifested in humans because of our upright posture.

Convection Requirement

The net rate of oxygen transport by the overall flow of air in and out of the lungs is the same as the oxygen transported by the overall blood flow through the lungs. This simple and, I hope, obvious statement has some important consequences. The relationship of the two bulk flows to oxygen transport can be described by the following equations: For ventilation,

$$\dot{V}_{O_2} = \dot{V}_{air}\beta_{gas}(P_{O_{2_{insp}}} - P_{O_{2_{exp}}})$$

and, for perfusion,

$$\dot{V}_{O_2} = \dot{Q}\,(C_{O_{2_{art}}} - C_{O_{2_{ven}}})$$

If capacitance and P_{O_2} are converted to concentration values and the two equations are combined then

$$\dot{V}_{air}(C_{O_{2_{insp}}} - C_{O_{2_{exp}}}) = \dot{Q}(C_{O_{2_{art}}} - C_{O_{2_{ven}}})$$

It now becomes clear the overall relationship of ventilation to perfusion depends on the respective concentration differences in the air and blood. A useful way to look at the concentration difference is through the concept of *extraction*, that is, the amount that is removed relative to the amount present. For air and blood, extraction (E) can be defined by the following two equations:

$$E_{air} = \frac{C_{O_{2_{insp}}} - C_{O_{2_{exp}}}}{C_{O_{2_{insp}}}}$$

and

$$E_{blood} = \frac{C_{O_{2_{art}}} - C_{O_{2_{ven}}}}{C_{O_{2_{art}}}}$$

In terrestrial mammals, extraction values for both blood and water are usually about 0.25 and tend not to vary much.

Some additional algebraic manipulations will allow us to come to a useful conclusion about ventilation and perfusion, so stick around to see how it all comes out. First, we will multiply the concentration differences by a fraction that equals 1.0, so that extraction can be substituted into the equations for $\dot{V}_{O_2}$. For convenience, I will show the substitution only in the case of air; the O_2 transport equation for blood works identically.

$$\frac{(C_{O_{2_{insp}}} - C_{O_{2_{exp}}})}{C_{O_{2_{insp}}}} C_{O_{2_{insp}}} = E_{air} C_{O_{2_{insp}}}$$

so the overall equation for O_2 transport becomes

$$\dot{V}_{O_2} = \dot{V}_{air} E_{air} C_{O_{2_{insp}}}$$

The relationship of the amount between the medium moved and the oxygen consumption is called the *convection requirement.* The time units of the fraction cancel, so the resultant dimensions are volume of medium (liter) over amount of oxygen consumed (mM). The above equation can be rearranged one step more to define convection requirement as follows

$$\frac{\dot{V}_{air}}{\dot{V}_{O_2}} E_{air} C_{O_{2_{insp}}} = 1.0$$

This format gives what is called mathematically a hyperbolic equation. The dimensions of the variables in the equation all cancel because the equation equals 1.0. Further, as any dimension increases, some other dimension must decrease proportionally. If we leave the extraction constant, the consequence is that, in the case of air, as inspired concentration is made smaller, the amount of air that must be moved through the lung must be greater. You may recall that a similar conclusion was reached in the initial discussion of capacitance. The argument in the case of convection requirement is really just another case of the same principle because concentration at a given P_{O_2} is a function of capacitance. The relationship between convection requirement and concentration of the medium coming in to the gas exchanger is given graphically in Fig. 6-1. Notice that the effect of a change in extraction is to move the curve slightly but that the overall relationship stays the same. In further confirmation of the earlier discussion of capacitance, note that an animal dependent on breathing water, where the concentration of O_2 is very low even

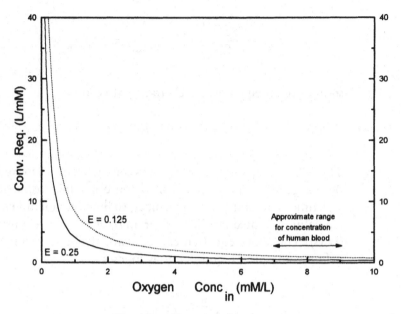

Fig. 6-1. Convection requirement.

when equilibrated with air, must move a very large amount of water over the exchanger to transfer a given amount of O_2.

If we compare the convection requirements of mammals for blood and air, we will find that, because the arterial O_2 concentration and inspired air concentrations of oxygen are nearly the same, 8–9 mM/L, about as much blood must pass through the lung as air as the two media simultaneously exchange gases. The overall conclusion to be drawn for humans is that, for the whole lung, we can expect the ventilation to perfusion ratio to be roughly one to one.

So long as the medium has a capacitance that is similar to the blood, there appears to be little selection pressure to improve the gas exchange strategy as, for example, might occur with a countercurrent exchanger. Birds in flight have high aerobic demands and may ascend to a high altitude where the concentration of O_2 in the air is much lower than at sea level. They have evolved a much more complicated unidirectional ventilatory system, perhaps to escape the otherwise high convection requirement that would be necessary with a constant pool exchanger.

Respiratory Exchange Ratio and R Lines

A further useful relationship can be derived from the capacitances of the respiratory media for oxygen and carbon dioxide and the fact that the exchange of both gases occurs simultaneously and at the same place. The ratio of the amount CO_2 produced to O_2 consumed is called the *respiratory exchange ratio (R_E)*. So,

$$\frac{\dot{M}_{CO_2}}{\dot{M}_{O_2}} = R_E$$

R_E is typically about 0.85 and may vary, depending on diet, from 0.7 to 1.0 (see Text Box 6-1 for more information on the factors that determine R_E).

The changes in partial pressure of the two gases during this gas exchange have a necessary ratio that is related to the capacitances and R_E. β_{gas} is the same for O_2 and CO_2, so, during simultaneous exchange in an aerial medium, it is reasonable to expect the respective changes in partial pressure to be the same as R_E. Of course, the inspired air has a P_{O_2} of about 20 kPa and a P_{CO_2} that is essentially zero so we can expect the exhaled partial pressures to change in about equal amounts, but in opposite directions. Therefore, in the exhaled gas, the P_{O_2} will be about 15 kPa and the P_{CO_2} will be about 5 kPa. The graphical representation of this relationship (Fig. 6-2) shows us that whatever the change in P_{O_2}, the change in P_{CO_2} will be linearly related to it as a function of the R_E and the relative capacitance of the medium for the gas. This relationship is called an *R line*. If one assumes that the ventilation of the lung has evolved to ensure adequate exchange of O_2, then it becomes apparent why there is such a

---

Text Box 6-1.   Diet and R_E.

The amount of O_2 consumed and CO_2 produced during the oxidation of common nutrients is determined by the chemical composition of the nutrient. Carbohydrates such as sugar or starch yield one molecule of CO_2 for every O_2 consumed, so the R_E for carbohydrate is 1.0. Oxidation of lipids yields less CO_2 and the R_E for an individual metabolizing lipid only is about 0.7. The R_E for protein depends on the composition of the specific amino acids, but for average dietary protein, an R_E of 0.8 is assumed. A mixture of these nutrients in the average diet results in the R_E of 0.85.

R_E is measured at the gas exchanger and not at the point of the actual metabolism. It represents the summed gas exchange of the body when it is measured. The gas exchange of cellular metabolism, called the *respiratory quotient* (RQ), and R_E are the same in a steady-state condition. Transiently, during hyperventilation, for example, R_E may differ substantially from RQ so the two terms are not necessarily synonymous.

---

high P_{CO_2} in the blood of an air-breathing animal and a comparatively low P_{CO_2} in the blood of a water-breathing animal. In air breathers CO_2 accumulates in the blood until the difference in P_{CO_2} between the lung and the air is essentially the same as the difference for P_{O_2}.

The low capacitance of water for oxygen requires so much ventilation of the gas exchanger that the P_{CO_2} falls to near equilibrium with the water. The R line for water breathers in Fig. 6-2 has, consequently, a much shallower slope than the R line for air breathers.

R lines for transport that involves diffusion rather than convection will have slightly different slope because of the difference in diffusion between O_2 and CO_2. In practice, this difference is small. The diffusion through the alveolocapillary membranes is followed by another bulk flow transport system, the blood. The capacitance of blood for CO_2 is greater than for O_2, so the slope

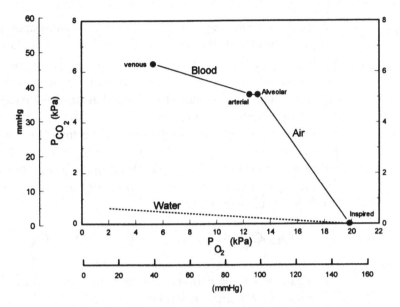

Fig. 6-2. R lines for air and water breathers.

of the R line in Fig. 6-2 for blood transport between the lung and the tissue has a shallower slope that is evidenced in the aerial phase of gas exchange.

Shunt Blood Flow

Despite the apparent overall adequacy of the ventilation and perfusion of the lung, measurements of the arterial P_{O_2} show it to be less than the alveolar P_{O_2}. This disparity can only mean that some of the blood from the pulmonary circulation is not exposed to gas exchange because the transit time through alveolar capillaries is known to be sufficient for diffusion equilibrium. There are blood vessels that do not go to the capillary circulation of the lung. The flow through these vessels is called *shunt* blood flow.

Knowledge of the concentration of O_2 in the blood entering and leaving the lung, together with the assumption that any blood leaving the alveolar capillaries is at a concentration appropriate to equilibration with the alveolar P_{O_2}, allows the calculation of the fraction of the blood flow that is apparently shunted. No blood leaves the system; it flows through shunts or through alveolar capillaries. Thus the O_2 added to the blood in the alveolar capillaries acts as a marker to quantify the flows. The O_2 transported from the lung is equal to the product of the total flow, that is, the cardiac output, Q, and the arterial O_2 content. The two constituents of this output are the O_2 transported from the shunt and the O_2 transported from the alveoli. Algebraically, this relationship can be described as follows,

$$Q_{total}C_{a_{O_2}} = Q_{shunt}C_{\bar{v}_{O_2}} + (Q_{total} - Q_{shunt})C_{A_{O_2}}$$

This equation can be rearranged to give the ratio of shunt blood flow to total blood flow as follows:

$$\frac{Q_{shunt}}{Q_{total}} = \frac{C_{A_{O_2}} - C_{a_{O_2}}}{C_{A_{O_2}} - C_{\bar{v}_{O_2}}}$$

The algebraic manipulation calculates the shunt blood flow as though any blood that might have been partially oxygenated were partitioned proportionally into either of the two categories. Of course, that is not what happens in the lung. The blood flow leaving the lung is the combination of flows from different regions of the lung, each of which may be differently oxygenated according to the adequacy and matching of its local perfusion and ventilation.

Specific Matching of Ventilation and Perfusion

The flows of the two respiratory media, air and blood, can match in three different ways. First, they can resemble closely the overall

$\dot{V}/\dot{Q}$ ratio, that is, about one to one. In this case the gas exchange results in the idealized complete equilibrium of partial pressures.

If, however, the perfusion is limited relative to the ventilation, then another condition obtains. In this second case, the blood will deliver rather less CO_2 to and take away rather less O_2 from the alveolar gas. As a result, the alveolar P_{CO_2} will fall and the P_{O_2} will rise. The blood flow leaving the capillaries in this case will have little effect on the overall gas transport because it is so small.

In the third case, the ventilation is limited relative to the blood flow. Here, the blood will not be able to get rid of its CO_2 as well or to pick up much O_2. As a result, both the gas in the alveolus and the blood leaving the capillaries will have a high P_{CO_2} and a low P_{O_2}.

The fluid pressures in the airways and pulmonary capillaries are similar in magnitude. Given the pliant structure of the exchange surface, any imbalance of the pressures will cause either the alveoli to expand and collapse the capillaries or vice versa. The upright posture of humans exacerbates this $\dot{V}/\dot{Q}$ mismatch because of the hydrostatic pressures in the column of blood from the apex of the lung to the diaphragm.

At the apex of the lung, only the systolic pulmonary pressure may be sufficient to force the capillaries open. Throughout the rest of the cardiac cycle while the alveoli are ventilated the blood flow may be stagnant because the pressure in the capillaries is insufficient to overcome the weight of the column of blood against which it works. Thus, the upper regions of the lung, sometimes called the *zone of collapse*, may have a high $\dot{V}/\dot{Q}$ ratio. The increase of pulmonary arterial pressure that comes with increased cardiac output can overcome the hydrostatic pressure of the blood and improve the $\dot{V}/\dot{Q}$ ratio. This change would be seen as in an increase in the diffusing capacity (D_L) of the lung because it would effectively increase the area of the exchange surface.

Moving from the apex towards the diaphragm the balance of pressures reverses. The $\dot{V}/\dot{Q}$ ratio changes to favor perfusion as the hydrostatic pressure changes from negative above the heart

Effects of regional pressure differences on V/Q

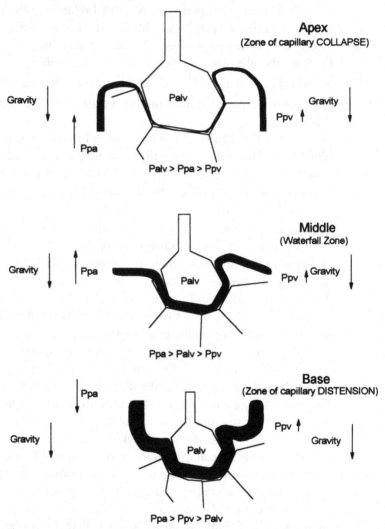

Apex
(Zone of capillary COLLAPSE)

Gravity

Palv

Gravity

Ppv ↑

Ppa

Palv > Ppa > Ppv

Middle
(Waterfall Zone)

Gravity   Ppa

Palv

Ppv Gravity

Ppa > Palv > Ppv

Base
(Zone of capillary DISTENSION)

Ppa

Ppv ↑

Gravity

Palv

Gravity

Ppa > Ppv > Palv

Fig. 6-3. Zones of the lung as defined by the relative pressures in the pulmonary artery (Ppa), exchange airways (Palv) and the pulmonary vein (Ppv). In the zone of collapse Palv > Ppa so blood flow may be stopped. In the middle (waterfall) zone Ppa > Palv > Ppv. In the zone of distension Ppa > Palv so ventilation may be impeded.

to positive below the heart, and the airways tend to be compressed as the blood vessels distend. Between this *zone of distention* and the zone of collapse, at about the level of the exit of the pulmonary artery from the heart, there is a balance of the pressures and ventilation and perfusion are most nearly balanced. This middle zone is picturesquely call the *waterfall* zone because all the pressures among the arterial blood, the exchange airways and the venous drainage change in descending order. These zones are illustrated in Fig. 6-3.

Many factors such as posture and forcefulness of breathing alter the relative size of these zones. Lying down will cause a reduction of both the upper and lower zones as the hydrostatic column is reduced substantially. Forceful exhalation or inhalation, particularly against a resistance, can substantially alter blood flow through the lung for a short time. The increase of airway pressure when one tries to exhale against a closed glottis, called a *Valsalva maneuver* can completely stop blood flow through the lung. The reduced return to the heart can have the dramatic effect of a momentary interruption of the heart beat.

Forceful inhalation can reduce the pressure in the lung and increase the blood volume of the lung enough to pull blood back into the lung from the pulmonary vein with the same effect on the heart. In both of these cases the effect is temporary and the heart compensates rapidly for the change in pressure.

CHAPTER 7

TRANSPORT OF GASES BETWEEN THE ALVEOLUS AND THE BLOOD

The gases within the exchange portion of the alveoli arrive there in part by bulk flow and in part by diffusion. Transfer across the alveolar surface from the alveolus to the blood occurs by diffusion alone. The barrier to this diffusion, the *alveolocapillary membrane* is thin but complex. After the nature of the barrier is given its obligatory description, we can safely consider it as a uniform structure for the time being. We will return to effects of variations in its composition later. For simplicity, this following discussion will emphasize the transport of oxygen from the alveolus to the blood. The transport of carbon dioxide across this barrier differs only slightly, other than in its direction.

The Diffusion Barrier

The barrier between the alveolar gases and the blood (Fig. 7-1) is comprised of the fluid lining the alveolar wall, the alveolar

75

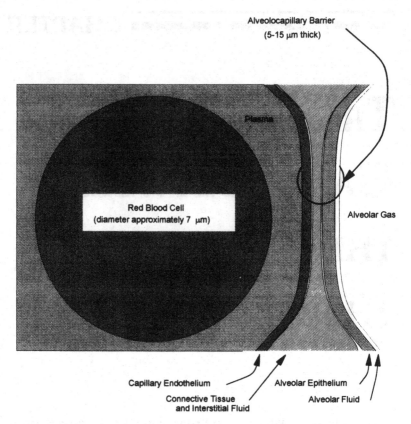

Fig. 7-1. The alveolocapillary diffusion barrier.

epithelium, interstitial fluid and connective tissue, and the capillary endothelium. This barrier is, appropriately for a good gas
exchange surface, quite thin despite its complexity. It is between
5 and 15 μm thick. The difference in P_{O_2} across the barrier is
so small that, for practical purposes, it is often considered to be
zero. (The small partial pressure difference is additional evidence
that the alveolocapillary membrane is a good diffusive gas exchange surface.) Having diffused across the tissue barrier, the
gases dissolve in the plasma. For gases that are also carried in
the red blood cells, there is the additional step of diffusion through
the plasma and the membrane of the red blood cell.

The arrangement of the capillaries between the alveolar walls resembles a meshwork rather than a set of parallel vessels. Because of the complexity of the possible paths for a given red blood cell, the time available for the transfer of gases can only be estimated statistically. For a given unit of blood in the alveolar capillaries, the average time passing through the exchanger is estimated to be about 0.75 s at rest. During activity, when the perfusion through the lung is greater, the blood may flow two to three times faster.

Figure 7-2 shows the times for equilibration of blood with different alveolar gases. The diffusion constant for oxygen is smaller than for carbon dioxide, so equilibrium between the oxygen partial pressures in the alveolus and blood is reached more slowly. Nevertheless, the equilibrium is reached well within the limits of the time available for exchange. The equilibrium for carbon dioxide is reached still more rapidly. This difference in equilibration times allows us to form a general rule about diffusive gas exchange in the body: If oxygen diffusion is sufficient, then

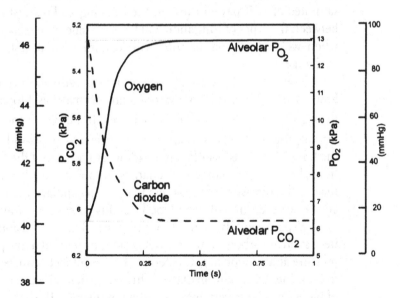

Fig. 7-2. Equilibration times for gases diffusing between the alveolus and the blood.

we will assume that the transfer of carbon dioxide, because of its more rapid diffusion, is also sufficient. This rule is another reason why we neglect much of the detail in CO_2 exchange. The study of oxygen diffusion is more likely to reveal the limiting factors of gas exchange.

Limitations of Gas Transfer

The average transit time exceeds the equilibration time, so diffusion through the alveolocapillary membrane is not what limits the exchange. This examination of the transport of O_2 to the blood suggests that more oxygen could be transported to the tissue if only there were more blood flowing through the alveolar capillaries to take it away. One may refer to this gas exchange in this case as being *perfusion limited* rather than *diffusion limited*. By our rule then, the exchange of CO_2 is also perfusion limited. If the exchange were diffusion limited, the blood would not be saturated after it passed through the exchanger. This distinction between the types of limitation will be of considerable importance when we look at how to cope with conditions that restrict gas exchange.

The exchange of O_2 can be considered perfusion limited so long as the P_{O_2} of the blood does not decrease as blood flow through the alveolar capillaries increases. If the P_{O_2} of the blood leaving the exchange surface were to decrease, it would mean the transit time was insufficient to allow saturation and the limitation of the gas exchange had changed to a diffusion limitation; that is, the diffusion of O_2 across the alveolocapillary membrane could not keep up with the blood's ability to carry it away.

Experimentally, a nonrespiratory gas that is diffusion limited, such as carbon monoxide (CO), administered at a low partial pressure that is not life threatening to the subject, can be used to examine the condition of the diffusion barrier. The capacitance of blood for CO is very high, so, at a low alveolar P_{CO}, the blood is nowhere near equilibrated with alveolar P_{CO} during its transit through the alveolar capillary. Its transport is therefore not, according to our definition, perfusion limited. Increasing the blood flow will not increase the transport of the gas. In this case, as in all cases of diffusion limitation, only the improvement of those variables that affect diffusion will improve the transfer of gas.

The causes of diffusion limitation are the insufficiency of the partial pressure difference as might occur in a hypoxic environment, an increased thickness of the diffusion barrier as might occur from accumulation of fluid in the interstitial spaces, or a decrease in exchange surface area as might occur from loss of lung tissue in emphysema. The diffusion constant of the lung itself, being rather like that of the its cells' principle component, water, is less likely to change. Irrespective of the specific cause, the simplest way to compensate for a decreased diffusion of O_2 is to increase the alveolar P_{O_2} by increasing the inspired P_{O_2} because manipulation of other diffusion variables is impractical.

If the transport of oxygen results from reduced carrying capacity of the blood as might occur from anemia, increased alveolar P_{O_2} will help very little. The hemoglobin, although reduced in concentration, will still be saturated. An increase in alveolar P_{O_2} will only increase the amount of oxygen dissolved in the water and that, as we have discussed earlier, is too little to help. Only a treatment that will increase the ability of the blood to carry O_2 away will be a successful treatment because the problem lies in the blood's capacity to carry O_2, not in the lung's permeability to it.

In summary, diffusion limitation and perfusion limitation differ operationally in terms of the factors that will affect gas exchange. When one wishes to alter gas exchange, it is important to employ a treatment that is appropriate to the circumstance. It will do no good to increase, say, the partial pressure difference across the barrier if the limitation to gas transfer is insufficient blood flow. It will do no good to increase blood flow if, for example, the thickness of the barrier is limiting the diffusion. Each condition can only be affected by a factor appropriate to it. As we will see in Chapter 9, this simple and logical dictum is not always followed in practice.

Lung Diffusing Capacity

The model that we have used to describe diffusive gas transport can only be directly applied to understanding the overall gas transport in the lung under two conditions: first, if all the variables of the diffusion equation were uniform throughout the lung or, second, if all the differing partial pressures, surface areas, membrane thicknesses, and diffusion constants were known for each part of the exchange surface. Clearly, neither of these conditions can be met. Further, it is not clear that it is useful or necessary to have such a rigorous analysis. A simpler model, one that describes the overall function of the lung as a diffusion exchanger, is more appropriate because only the whole lung can be treated in an intact subject.

The diffusion constant, as we have discovered earlier, is already comprised of several not clearly intuitive dimensions. If, for purposes of the description of the whole lung, we incorporate into the diffusion constant those factors we cannot practicably measure, there results a new term, *lung diffusing capacity*, that is defined using only the conveniently measurable variables. These variables are the amount of gas exchanged and the overall partial pressure difference across the exchanger.

Where $\dot{V}_{gas}$ is the rate of gas transfer, ΔP is the partial pressure difference across the exchange surface, and D_L is the lung diffusing capacity, we can rewrite the diffusion equation as

$$\dot{V}_{gas} = D_L \Delta P_{gas}$$

and, solving to define D_L, we have

$$D_L = \frac{\dot{V}_{gas}}{\Delta P_{gas}}$$

If D_L were measured using oxygen, it would require knowledge of the mixed venous P_{O_2} and the alveolar P_{O_2} would have to be reduced to a low enough value to make its transport diffusion limited. If low levels of CO are used, the technique becomes much simpler. First, CO is known to be diffusion limited and, second, we can assume its initial venous partial pressure to be very nearly zero. In this case, the equation simplifies to

$$D_L = \frac{\dot{V}_{CO}}{P_{A_{CO}}}$$

The value for D_L obtained by this method can be used directly for the understanding of lung function in an intact subject by comparing its change under different conditions such as rest and exercise. Because each individual differs somewhat in the variables involved in diffusive gas exchange, the use of lung diffusing capacity for comparisons between individuals is less valid.

It should be obvious that any of the factors that affect diffusion will similarly affect the diffusing capacity of the lung. In reality, there are additional factors that come into play. Factors such as posture, exercise, and disease will manifest themselves in changes of D_L. Chief among these is the local relationship of ventilation to perfusion in different regions of the lung. This matching, as we discussed in the previous chapter, is often not uniform. Much greater consistency has been found using *specific diffusing capacity*, the ratio of D_L to V_A (Rosenberg et al., 1986). This value is independent of lung volume and posture and may be the basis of a more reliable standard.

CHAPTER 8

TRANSPORT OF OXYGEN IN THE BLOOD

As we saw in the earlier chapters, the capacitances of water and air for respiratory gases impose some rigorous constraints on the environmental side gas exchangers. The differing physical constraints within a closed circulatory system provide yet another set of limitations. Although blood is much more complex (in addition to being thicker) than water, the means by which oxygen is carried in it nevertheless operate on the same set of fundamental principles of gas transport: diffusion, bulk flow, and capacitance.

Capacitance of the Blood for Oxygen

As has been the case in previous discussions, I will temporarily ignore CO_2 to concentrate on O_2. CO_2 is much more interesting in its relationship to respiratory acid-base balance and will be discussed in context with that topic.

Refer to Fig. 8-1 as a visual aid in the discussion of O_2 transport. It has been extracted from Fig. 1-3 to emphasize the

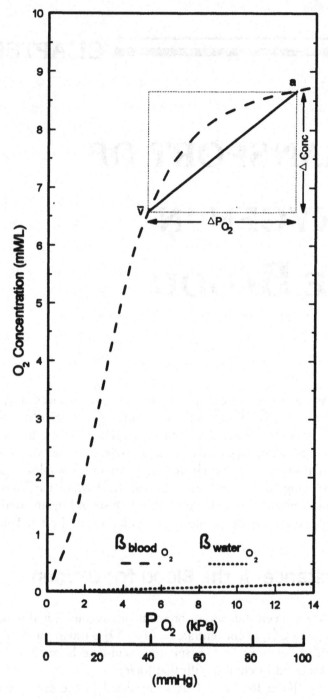

Fig. 8-1. Blood-Oxygen equilibrium and water-oxygen capacitance.

comparison of the interaction of O_2 with blood and with water. A simple calculation based on the data that are shown in Fig. 8-1 can be instructive in the hypothetical case that O_2 were carried only as an aqueous solution in the plasma. We can measure the difference in arterial and mixed venous P_{O_2} as about 7.5 kPa in humans. (These and the following SI units are converted to units that may be more familiar to some readers in Text Box 8-1). The solutes in plasma would lower its capacitance below water, but the value for pure water is close enough for these purposes. β_{waterO_2} at 37°C is 10.5 μM/(L × kPa) so the change in concentration between arterial and venous plasma is a little less than 80 μM/L. With a cardiac output of about 5 L/min, the rate of oxygen transport would be about 0.4 mM/min.

But the O_2 consumption is known to be about 9 mM/min, much less than our calculation indicates. To increase the bulk flow of the blood to carry enough O_2 in solution in water, the cardiac output *at rest* would have to be more than 20 times greater, over 100 L/min. Clearly, a simple increase in the flow rate is an impractical solution to the problem of increasing O_2 transport in the blood. The more workable solution was apparently to evolve a means to make the capacitance of blood 20 times greater than water, instead of increasing the flow rate.

Text Box 8-1. O_2 Content of the Blood: Conventional Units

In conventional units, the change in P_{O_2} between arterial and venous blood is about 55 mmHg. If the blood were water the resulting change in concentration would be about 0.18 ml O_2/100 ml and the amount transported at a cardiac output of 5 L/min would be about 9 ml O_2/min. Normal oxygen consumption is about 200 ml/min.

Human blood has a capacity for oxygen of about 20 ml O_2/100 ml and contains about 15 g Hb/100 ml. Neglecting the small amount dissolved in water, the combining capacity of hemoglobin is thus 1.34 ml O_2/g in these units.

Hemoglobin

Throughout the biological realm there are many molecules that are involved with transport of oxygen. Many of these have in them a lattice structure, a porphyrin ring, that holds a transition metal atom that, in turn, combines reversibly with an oxygen molecule. In aqueous solutions these molecules are often distinctly colored, thus, they are called *respiratory pigments*. Most vertebrates use the respiratory pigment *hemoglobin* (Fig. 8-2), a molecule with a mass of about 68 kDa in which the O_2-combining or *heme* portion has an iron atom in its porphyrin ring. For brevity, the symbol Hb is often used to represent hemoglobin.

There are four heme binding sites in the molecule so, at saturation, a hemoglobin molecule can combine with exactly four O_2 molecules. The hemes are surrounded by and held in position by four proteins that comprise the *globin* portion of hemoglobin. As can be seen in Fig. 8-2, Hb combines with compounds other than O_2. Binding with CO_2 and DPG (diphosphoglycerate, now more correctly called bis-phosphoglycerate) affects the three-dimensional structure of the Hb molecule and, even though these molecules do not compete with O_2 for the same binding sites, they affect the affinity of Hb for O_2 and ions such as H^+ and Cl^-. There may be more than one kind of Hb present in an animal's blood, either simultaneously or at different times in its life. Substitutions of even a few amino acids in the proteins may also drastically alter the affinity of a particular hemoglobin for oxygen.

Hemoglobin is water soluble and its presence in vertebrate blood raises the $\beta_{blood_{O_2}}$ as a direct function of its concentration in the blood. A gram of Hb will take up about 0.06 mM O_2/g at saturation. Human blood combines with about 9 mM O_2/L at normal alveolar P_{O_2} and is essentially saturated, so it works out that there must be about 150 g of Hb in a liter of blood. Some diving animals that must store O_2 in their blood may have

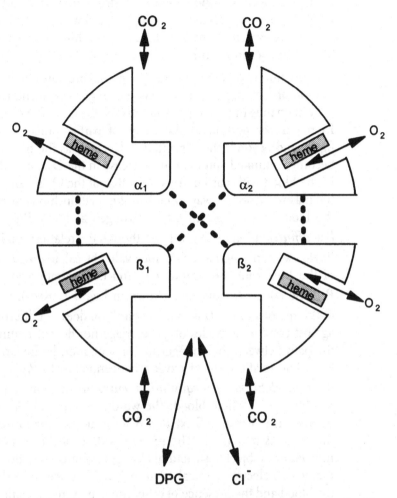

Fig. 8-2. Diagrammatic representation of the hemoglobin molecule. The proteins comprising the globin portion are labeled α_1, α_2, β_1 and β_2. The dashed lines between the proteins represent the various forces the bind them together. The sites for combination of O_2, CO_2, and DPG are indicated here to emphasize that these entities do not compete for the same binding sites. The effects CO_2, Cl^-, and DPG will be in the following chapters. This diagram was adapted from Imai (1982) and Voet and Voet (1990).

a carrying capacity of more than 13 mM/L or more than 200 g Hb/ L. Nondivers and less active animals have lower concentrations.

As is evident from Fig. 8-1, the combination of O_2 with blood is not a simple linear function of P_{O_2}. The $\beta_{blood_{O_2}}$ changes as a function of partial pressure. For partial pressures in the lower portion of the range, there is a large change in concentration for a given change in partial pressure; that is, the capacitance is high. In the upper portion of the range of partial pressures in the figure and beyond, the capacitance decreases as the hemoglobin becomes saturated and the only increase in the O_2 concentration in the blood with partial pressure is that of the O_2 in solution in the plasma. Another way to look at the relationship is to say that the blood has a *high affinity* for oxygen at lower P_{O_2}'s and a *low affinity* at higher P_{O_2}'s. Even though the relationship is non-linear, we can assign a capacitance value by calculating the slope of the line from the venous P_{O_2} and concentration to arterial P_{O_2} and concentration, as is shown in Fig. 8-1. Recall that one of the reasons to use β is that its definition does not discriminate against combined modes of gas carriage nor does it require that the path between the changes in concentration be linear.

The relationship of oxygen concentration to P_{O_2} is called the *hemoglobin-oxygen equilibrium curve* or, more appropriately for the case of whole blood, the word *blood* should replace Hb in the term. Refer to Text Box 8-2 for an explanation of how the curve is measured. The protein portion of the hemoglobin molecule is subject to structural changes that are a function of the molecule's electrochemical environment. The acid-base status of the blood and the presence of other molecules in association with hemoglobin can affect the readiness with which O_2 binds with the heme.

A single Hb molecule would have a stepwise equilibrium curve as each of its four sites accepted an O_2 molecule. The curves such as that in Fig. 8-1 represent billions of molecules, each of which is in its own peculiar situation that may differ from others. The result is that the curve becomes smoothed when it represents such a huge number of cells.

Text Box 8-2. Measurement of the Hb-O₂ Equilibirum Curve

The apparent color of hemoglobin changes as a function of its combination with oxygen. Most people are aware that venous blood is dark and arterial blood, or that exposed to air, is bright red. Physically, this change in color is measured as a change in absorption of light at a given wavelength. That change is directly proportional to the percent saturation of the Hb with O_2.

Measurement of the relative absorption of light by samples of blood that have been equilibrated with various partial pressures of oxygen allows one to plot percent saturation against P_{O_2}. A second determination, that of O_2 content at saturation, is necessary to convert the percent saturation data to actual O_2 concentration, as has been done in Fig. 8-1.

If one is dealing with samples of relatively uniform Hb content and interested primarily in changes in affinity, the use of percent saturation is usually sufficient. As we shall see, there are some pitfalls to that approach.

At partial pressures near zero, where the first O_2 molecules would be combining with otherwise unbound hemes, the affinity of the molecule is relatively low. Once that first association has occurred, there is an important change in the affinity of Hb for O_2. The presence of a combined O_2-heme group increases the affinity of the remaining uncombined hemes for O_2 at a given P_{O_2} and additional combinations are formed more readily. When most of the sites are saturated, the affinity again decreases. This saturation-dependent affinity gives the Hb-O_2 curve its characteristic S-shaped or *sigmoid* configuration. Just to increase the capacitance of the blood, it would be sufficient if concentration of O_2 in the Hb-O_2 relationship were a linear function of P_{O_2}. The sigmoid configuration gives the carriage of oxygen by the blood some additional and very useful features beyond the simple increase its capacitance.

Mixing of inspired and residual gases in the lung results in an alveolar P_{O_2} of about 13 kPa (100 mmHg) if the inspired air has the usual, near-sea-level P_{O_2} of about 20 kPa (150 mmHg). Conveniently, the blood is essentially saturated with oxygen at 13 kPa. But observation of the Hb-O_2 curve shows us that the blood will be nearly as saturated at partial pressures of oxygen below 13 kPa. The blood is still about 75% saturated at a P_{O_2} half that found in the alveolus.

The immediate benefit of this relationship is that it allows the blood to be oxygenated at partial pressures of oxygen well below that we might normally encounter. That this phenomenon provides a seeming safety factor is reassuring to engineers but it is curious to biologists who are accustomed to the evolution of traits that are "just good enough." What selective pressures would lead to such an enormous overcapacity for saturation?

It is obvious that, with this trait, animals can invade hypoxic atmospheres such as high altitude. That is a good fortune for mountain climbers and those few species that live at high altitudes, but most species, humans included, evolved and live at altitudes that are near sea level and where the oxygen partial pressure is never low. Environmental hypoxia cannot have been much of a selective pressure. I think I know what has gone on here and we will go into it in Chapter 9, but there are other pieces of the picture to be assembled before we go in that direction.

The Concept of P_{50}

Among different species and under changing conditions within a given individual, there are many factors that can and do affect the shape and position of the Hb–O_2 equilibrium curve. Changes in the curve reflect changes in the affinity of blood for oxygen. The affinity of blood for oxygen can be described mathematically to allow a more quantitative analysis of its changes. However, the equations that describe sigmoid curves are too unwieldy for convenient use. A simpler way to describe the differences in the various curves is to choose a characteristic point on the curve as its identifier. In the case of the sigmoid Hb–O_2 curve, that point lies in the midst of a curve's steepest slope, that is, its greatest capacitance. It is called the P_{50} and is defined as the partial pressure of oxygen at which the blood in question is 50% saturated. Any change in the affinity of blood for oxygen can be readily characterized by the change in the P_{50} of the curve. Refer to Fig. 8-3 to see how the P_{50} clearly describes different curves.

In the case of human blood, the P_{50} is about 3.4 kPa (25 mmHg). When the P_{50} increases, it means that the blood's affinity for oxygen has decreased because a greater partial pressure is required to achieve half-saturation. Conversely, if the P_{50} were decreased, it would mean the blood became more saturated at a lower P_{O_2}, that is, its affinity for O_2 had increased.

The tissues have P_{O_2}'s that vary with their activity and perfu-

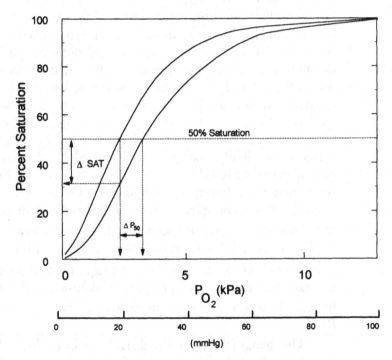

Fig. 8-3. The effect of changes of P_{50} on unloading of O_2 from blood. The dashed lines illustrate the two ideal cases discussed in the text.

sion and the P_{O_2}'s of the venous blood leaving them varies accordingly. Typically, the mixed venous blood returning from the tissues has a partial pressure that is higher than the P_{50}. Thus, on average, less than half of the O_2 carried in the arterial blood is consumed by the tissues. As demand for oxygen increases, the saturation and P_{O_2} become lower. Intense muscular activity can reduce the venous blood oxygen content in the vessels leaving those tissues to near zero.

Effects of Decreased Affinity of Blood for Oxygen

Tissues can only get oxygen from the blood by having a lower P_{O_2} than the blood. But, the lower P_{O_2} at the cell surface limits the subsequent diffusion gradient to the mitochondria within the cell. The cells can maintain their internal diffusion gradient if the affinity of blood for oxygen is less; that is, the blood has a higher P_{50}. At a higher P_{50} the blood will be less saturated and more oxygen will have been made available to the tissue without sacrifice of the gradient from the cell surface to the mitochondria.

Another way to look at this phenomenon is to observe that a change in affinity that gives an increased P_{50} decreases the capacitance of the blood for O_2; the slope of the blood-O_2 equilibrium curve is shallower, meaning that less O_2 can be held at a given P_{O_2}. Capacitance has been discussed heretofore in the context of loading more O_2. It is important to recognize that decreasing $\beta_{blood_{O_2}}$ is useful as O_2 is unloaded. In this case, it results in more O_2 delivery to the tissue without requiring that the tissues decrease their own P_{O_2} and compromise diffusion gradients into the cell. In simpler terms, an increased P_{50} "defends" the tissue P_{O_2}.

The shape of the Hb molecule is altered by its electrochemical environment. In the presence of increased levels of CO_2 or acidity, the affinity of Hb for O_2 is decreased because the globin portions of the molecule alter their conformation slightly and access to the heme groups is changed. The change in P_{50} under these conditions is called the *Bohr shift* or *Bohr effect*, after its discover (Bohr et al. 1904). Figure 8-4 has been drawn from their original data. Note again how easily the change in the shape of the curves is characterized by the use of the change in P_{50}. The P_{50} is also increased by increased temperature. Working muscles, for example, can be a few degrees warmer than the core body temperature. The resulting increase in P_{50} as the blood flowing into the muscle capillary beds is warmed will mean that the saturation of the blood will be less at a given tissue P_{O_2}.

The consequence of the Bohr and temperature effects on

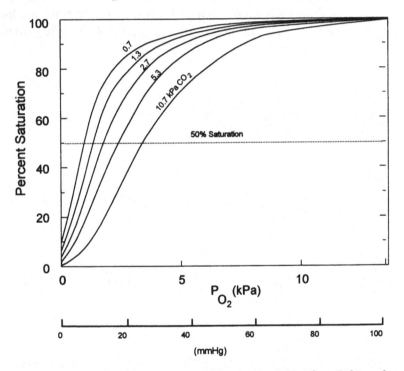

Fig 8-4.   Effect of P_{CO_2} on the O_2 equilibrium curve. Data from Bohr *et al.* (1904) converted and redrawn in modern units. The experimental P_{CO_2}'s used by the authors were in mmHg and have been rounded to the nearest tenth kPa.

the P_{50} is that oxygen will be more readily given up where the demand is higher. When the O_2-depleted and CO_2-loaded blood returns to the lung, the Bohr effect works in the opposite way. As CO_2 leaves the blood, the P_{50} decreases signifying an increased affinity for O_2. This change increases the capacitance and the blood thus will carry an increased amount of O_2 at a given P_{O_2}. It is important to notice that the P_{50} changes as the blood transits through the circulation. The equilibrium curves drawn with variables other than P_{O_2} held constant never occur in life. There is, in realistic terms, a dynamic equilibrium "loop" or hysteresis curve that reflects the changes in P_{O_2} and oxygen content as a function of the location in the circulatory system. It is not possible to draw such a loop quantitatively because the time and location of a "particle" of blood varies so greatly at any given moment.

CHAPTER 9

Oxygen Transport in Hypoxic Conditions

When the O_2 supply to the mitochondria fails to keep up with the demand, the tissue P_{O_2} falls and the tissues are said to be *hypoxic*. Hypoxia is the most significant stress to which the respiratory system must respond. To be sure, there are conditions wherein the transport of CO_2 is critical, but, in general, O_2 transport problems are the more common occurrence. The cause of the hypoxia can be at any part of the transport of oxygen or from the lack of sufficient O_2 in the environment. The types of hypoxia are often named in accordance with the cause. Text Box 9-1 lists common hypoxias. To illustrate how the respiratory system copes with hypoxia, I will discuss four circumstances in which the body must respond to hypoxic conditions: high altitude, pregnancy, exercise, and disease.

In all types of hypoxia, the tissues will be under the same stress, and the short- and long-term means with which the body responds to the stress will be fundamentally the same, irrespective of the cause. Changes in the fundamental operating principles of gas transport, capacitance, diffusion, and bulk flow are the

Text Box 9-1.   Types of Hypoxias

If the P_{O_2} of the inspired air is low enough to result in reduced loading of the blood with oxygen, the condition is said to be *hypoxic hypoxia*. In accordance with the earlier discussion, the result is a diffusion limitation and can be compensated only by alterations of the factors that affect diffusion.

Anemic hypoxia results from an insufficient capacity of the blood to carry oxygen. In this case, there is a perfusion limitation and increases in diffusion-related factors will usually avail little. Increased blood flow rate or, more importantly, increases in the blood's ability to carry O_2, such as by increasing the Hb concentration, will be most effective in compensation for this perfusion limitation.

If the perfusion of the tissue capillaries is insufficient, we have another kind of perfusion-limited gas exchange called *circulatory hypoxia*.

Finally, the limitation may occur in the tissues or cells themselves, so that, despite an adequate supply of oxygen, its utilization is impaired. This *histotoxic hypoxia* is best exemplified by the example of cyanide poisoning in which the use of O_2 in the electron transport system of the cell is prevented.

only means available to the system, so it will not astound the reader to find we examine those factors.

High Altitude

Most humans and, for that matter, most animals do not live at altitudes where the barometric pressure lowers the P_{O_2} enough to have a significant effect on the diffusion of oxygen from the lung to the blood. With the possible exception of some longtime residents of the Himalayan highlands, there seems to be little evidence of a genetic human adaptation to high-altitude hypoxia. Nevertheless, responses to this form of hypoxic hypoxia have been studied most extensively and comprise the most commonly used examples. Perhaps this attention is paid because it is a form of respiratory stress that is readily grasped by students or it has had ,practical applications to aviators or just because it provides justification for the expeditions of mountain-climbing physiologists. For any of these reasons, it is worthy of discussion.

The acute ventilatory response to hypoxia is hyperventilation. An increase in alveolar ventilation, even at low ambient P_{O_2}, increases alveolar P_{O_2} and improves the diffusion gradient across the alveolocapillary barrier. The negative side of this response is that it results in lowering the blood P_{CO_2} and subsequent acid-base disturbances.

In the acute response, an increased heart rate, and, consequently, an increased cardiac output is observed. Increased perfusion may increase the diffusion gradient by removing oxygen more rapidly and, thus, transport more oxygen. The increased work the heart must do to increase cardiac output, of course, contributes to the problem by increasing demand.

Chronic hypoxia elicits a more complicated response that involves the blood. The hypoxia can be relieved by an increase of alveolar P_{O_2}, which would suggest a diffusion limitation even though the blood is equilibrated with the P_{O_2} in the alveoli. The test of diffusion limitation, however, is not the achievement of equilibrium, but whether increasing the potential for diffusion increases transport. In this case it does because the blood is not saturated with oxygen at the low P_{O_2}.

The increase in alveolar P_{O_2} that comes from hyperventilation is of limited help because the P_{O_2} of the incoming air is already low and the cost of increased ventilation exacerbates the problem. The decreased P_{CO_2} from hyperventilation is of little consequence itself and the kidney can compensate for the acid–base disturbance over a period of days. Chronic compensation for hypoxia requires that the blood be able to carry enough oxygen at the available alveolar P_{O_2} and deliver it to the tissues. Something in the blood must change.

Diphosphoglycerate, like other organophosphates, alters the

shape of the Hb molecule (refer to Fig. 8-2) and lowers the affinity for O_2, that is, increases the P_{50} of the Hb–O_2 equilibrium curve (Fig. 9-1). In humans, the concentration of DPG in the red cells increases in response to hypoxia. The effect of a lowered affinity is an increase the partial pressure at which the blood gives up oxygen. The increase of O_2 unloading can maintain the tissue P_{O_2} at near normoxic levels, so this seems a reasonable response. However, in the lung the lowered affinity results in less saturation of the Hb with O_2 so one might question, as some have, its appropriateness. In fact, some species of mammals that have apparently adapted evolutionarily to life at high altitude have just the opposite adaptation. They have Hb–O_2 equilibrium curves with a lower P_{50} (Fig. 9-1), that is, a higher affinity for oxygen. They can saturate their Hb with O_2 more completely in an hypoxic atmosphere but must then have tissues that function well at a lower P_{O_2} than most humans can tolerate to get the oxygen from the blood. This perplexing situation requires some further examination.

Note that the vertical axis in Fig. 9-1 is in units of percent saturation. In this form of the Hb–O_2 equilibrium graph, changes of the position of the curve are typically described as "right-shifted" or "left-shifted" from normal. Although it is in common usage, this terminology causes an important misunderstanding of the nature of the response and I would argue for its abolition in favor of description of the change in affinity in terms of P_{50}. The basis for my argument follows.

Representation of the response in terms of percent saturation is convenient but it omits what is probably the most important response to hypoxia whereas it emphasizes a less important response. With chronic exposure to hypoxia, cells in the kidney secrete the hormone *erythropoietin* that stimulates the production of red blood cells. These additional cells substantially increase the concentration of Hb in the blood and, therefore, its O_2 carrying capacity. This *polycythemia* has been questioned by some because, at low P_{O_2}, the blood cannot be saturated and it increases the viscosity of the blood and the work of the heart.

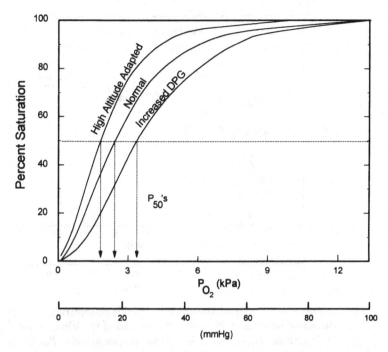

Figure 9-1. Changes of P_{50} of animals evolutionarily adapted to high altitude (fetal blood may be included in this category) and of those that acclimatized to this form of hypoxia by increasing DPG concentrations in the red blood cells. Because the vertical axis of this graph is in terms of percent of maximum saturation, the effect and importance of increased hematocrit is hidden.

Examination of Figs. 9-1 and 9-2 should clarify the effect of the response. The vertical axis of Fig. 9-2 is not percent saturation but, rather, O_2 concentration. This representation of human acclimatization includes not only the change in P_{50} but also the change in the O_2-carrying capacity of the blood. First, consider the position of the curve for the altitude-acclimated blood relative to the normal, low-altitude, curve. Although described as "right-shifted" its position is to the left of the normal curve! The P_{50} of the high altitude curve is greater than the normal curve in either representation and is therefore the more consistent way to represent the phenomenon.

The increase in P_{50} does make unloading to the tissues

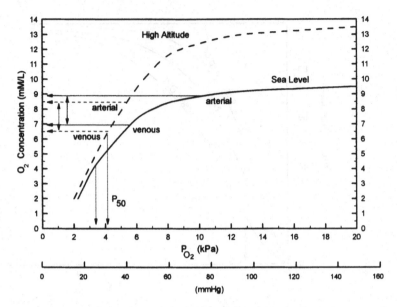

Figure 9-2. Acclimatization to high altitude emphasizing the importance of increased hematocrit. Note that the amount of O_2 delivered and the venous that is, tissue, P_{O_2} change little and that, despite its higher P_{50}, the dissociation curve after acclimatization, lies to the left of the sea-level curve.

possible without substantial lowering of the tissue P_{O_2}. The most important compensation comes from the increase in red blood cell concentration. Comparison of the difference in concentration, invisible on a percent saturation format graph, shows that the effect of increased red blood cell concentration lies not in the maximum carrying capacity but in the increased capacitance for oxygen that occurs in the range of partial pressures between loading and unloading. As a result of the greater $\beta_{blood_{O_2}}$ which is also obscured if percent saturation is used, the change in concentration between arterial and venous blood (the amount of O_2 delivered) is the same or greater at high altitude as at low altitude before the compensation.

The blood leaves the alveolar circulation carrying less O_2 than it might if it were equilibrated with an alveolar P_{O_2} appropriate to a low altitude, but the desaturation is irrelevant to the compensation. An increase in blood flow through the exchanger may not improve gas exchange very much but an increase of alveolar P_{O_2} will, so the exchange is diffusion limited. The diffusion from the capillaries to the tissue is not limited after compensation. Figure 9-3 shows the R-lines for normal and high-altitude-compensated gas exchange. Note that, because of the decreased affinity combined with the increased capacitance, the venous, that is, tissue, P_{O_2} is barely changed from normal, although the arterial P_{O_2} is much decreased.

Both the increased carrying capacity of the blood and the decrease in its affinity for oxygen are important to compensate for hypoxic hypoxia; neither does enough by itself. But together, their combined effects allow sea-level organisms to exist at high altitude with nearly normal oxygen delivery, in terms of amount and partial pressure, to the tissues.

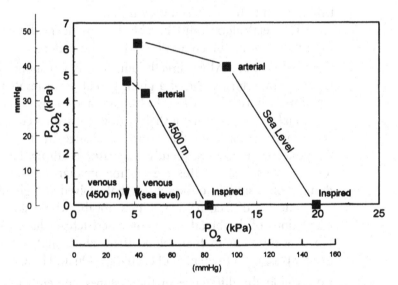

Figure 9-3. Acclimatization to high altitude represented with an R-line diagram. Data are from humans native to 4500m and at sea level. Note again that the tissue P_{O_2} remains high at high altitude because of the increased capacitance of the blood for O_2. The data have been recalculated from Torrance et al., (1970/71).

Pregnancy

The mammalian fetus lives in an environment of chronic hypoxic hypoxia. The necessity of diffusive exchange through the placenta means that its cascade of transport to the cells must begin with a P_{O_2} in the umbilical vein that is lower than the mother's arterial P_{O_2}. The placenta has an active metabolism of its own and extracts some of the O_2 delivered in the maternal blood before it is exposed to the blood of the umbilical artery. As a result, the blood in the umbilical vein has a P_{O_2} as low as 3 kPa in humans. In other mammals the P_{O_2} of the umbilical vein may be higher (Text Box 9-2). In those species such as humans, that provide the fetus with relatively hypoxic blood, we should expect that there will be adaptations in the gas transport of the fetus' respiratory system, especially in the blood, that allow the fetus to live and grow in this apparently hostile respiratory milieu.

The fetal blood must be able to saturate readily with O_2 from the maternal blood at a P_{O_2} where the maternal blood will readily desaturate. More directly put, fetal blood should have a higher affinity for oxygen at a given partial pressure; its $Hb-O_2$ curve must have a lower P_{50}. In humans and many other mammals fetal hemoglobin has a slightly different sequence of amino acids in the β portions that causes a change in the shape of the molecule and, consequently, increases its affinity for oxygen (decreases its P_{50}). In this regard, mammalian fetuses can be thought of as similar to the species that evolved at high altitude.

The P_{50} of fetal hemoglobin is not the whole story of the adaptations of the blood. In many cases, fetuses have a higher Hb concentration in their blood than adults that results in a greater $\beta_{blood_{O_2}}$ in the region of exchange. On an $Hb-O_2$ equilibrium graph the difference in these slopes (remember β is the slope of a graph of concentration and partial pressure) shows that, even with the same P_{50}, fetal blood will have a higher concentration of O_2 than maternal blood. Again, the responses of the fetus to its hypoxic environment is much like the response of an individual to high altitude. Of course, a pregnant mother

Text Box 9-2. Efficacy Ratios of Placental Gas Exchange

In Ch. 3 the concept of the efficacy ratio of gas exchangers was introduced as a means to compare the types of gas exchangers. Using data cited by Metcalfe et al. (1967), one can examine the differences in a few species of mammal's placental gas exchange. In this case the efficacy is defined by the version of the equation from Text Box 3-2 given below in which the P_{O_2} of the media involved is represented with the subscripts a and v for arterial and venous and um and ut for umbilical and uterine. (Recall that the umbilical artery, not the vein, carries O_2-depleted blood to the placenta.)

$$\text{Efficacy} = \frac{P_{v_{ut}} - P_{v_{um}}}{P_{a_{ut}} - P_{a_{um}}}$$

The ratios are calculated as follows.

Animal	Efficacy
Human	0.32
Monkey	0.31
Goat	0.32
Rabbit	−0.34

Note that there is a negative sign in the case of the rabbit, which suggests that the rabbit placenta must have a countercurrent exchanger. The P_{O_2} of the blood in the umbilical vein is higher than the uterine vein only in the rabbit. The other species may have countercurrent exchangers but are apparently not getting the full benefit of the design because the umbilical blood has a lower P_{O_2} than the uterine blood leaving the exchanger.

who goes to a high altitude presents her fetus with an extremely hypoxic stress.

The hypoxia that the fetal cells must endure is extreme in comparison to adult tissues. Recall that the fetal arterial (umbilical vein) P_{O_2} is only (amazing to me) 3 kPa. To extract oxygen from the blood, the tissue must be operating at a partial pressure near zero. Fetal tissue must develop and grow at a P_{O_2} that would render an adult unconscious or dead. Like the high-altitude-acclimated animal, the fetus can deliver a greater amount of O_2 for a given change in partial pressure. The $\beta_{blood_{O_2}}$ in the region of exchange is much larger because of its lower P_{50} and higher Hb concentration.

Fetal blood arrives at the placenta deficient in O_2 but also carrying CO_2. That means that its P_{50} is high because of the Bohr effect. Meanwhile, the maternal blood, lower in CO_2 has a low P_{50} because it is not carrying as much CO_2. As the exchange of CO_2 occurs, the two bloodstreams experience Bohr shifts in opposite directions. The affinity of the fetal blood increases while the affinity of the maternal blood decreases. Even if the Hb–O_2 curves of both were identical before exchange, they move apart, favoring exchange of O_2 from the mother to the fetus, as CO_2 is exchanged in the opposite direction. In species that lack fetal Hb and in experiments in which the fetal Hb has been exchanged by *in utero* transfusion for adult Hb, development of the fetus proceeds normally. Apparently, the opposite Bohr shifts and/or the difference in β are sufficient to allow the appropriate exchange of O_2.

At birth, the young mammal must drastically alter its gas exchange. Circulation that has barely perfused the lungs in favor of the placenta must alter its pattern and the lungs must inflate with air. The fetal blood, and consequently its tissues, will then be exposed to a theoretically instantaneous and large increase in P_{O_2}. In reality, the change in a newborn's arterial P_{O_2} takes hours to days to reach adult levels. The perfusion and ventilation of the lung are initially poorly developed and the infant's lung diffusing capacity is low at first. Fetal Hb is replaced by adult Hb and the tissue P_{O_2} rises slowly.

Disease

Disease conditions that limit gas exchange must be met with the same kinds of responses that are employed to cope with environmentally-imposed limitations. *Emphysema* is a destruction of lung tissue that reduces the exchange surface area of the lung. The gas exchange capability of the normal lung is far in excess of the resting demands so, in sedentary individuals, the disease, often caused by smoking, can progress all but unnoticed until the resting gas exchange is compromised.

As might be expected, the reduction in exchange area imposes a diffusion limitation. Increases in perfusion of the exchanger cannot help; the response must be to augment factors that are found in the diffusion model. The most immediate and obvious response is to increase the P_{O_2} in the lung with hyperventilation that, as we know, will increase the alveolar P_{O_2}.

A further insidious effect of the disease is that the connective tissue of the lung is also destroyed. The compliance of the lung is increased, making inhalation easier but the radial traction that keeps the airways open during exhalation is much reduced and, on exhalation, the airways collapse obstructing deflation of the lung. Individuals with emphysema characteristically have abnormally large residual volumes and tidal ventilation that goes on at near vital capacity.

As the disease progresses, supplemental O_2 is required to augment reduced P_{O_2} that comes from obstructed ventilation and reduced exchange surface. At first, any exertion by the afflicted individual will bring on hypoxic collapse. Subsequently, even the resting gas exchange cannot be met.

In cases of *pneumonia* or *pulmonary edema*, fluid accumulates in the alveoli and another case of diffusion limitation occurs because the distance between the gas phase and the blood is increased. Again, an increase in alveolar P_{O_2} is a reasonable way to deal with this hypoxia because it attacks the problem of diffusion.

Carbon monoxide poisoning occurs because Hb has a much higher affinity for CO than O_2. Even at low levels of CO, the carrying capacity of the blood for O_2 can be reduced effectively to that of the plasma solubility, that is, almost nil. The P_{O_2} of the arterial blood will be normal or even higher because there is no limitation of diffusion.

In this case of perfusion limitation, the necessary increase in perfusion, as successful suicides testify, is well beyond the capability of the heart to increase cardiac output. Treatment may include administration of high P_{O_2} gas mixtures, not to alter diffusion but in the hope of improving competition of O_2 for the binding sites occupied by CO and increasing, albeit slightly, the concentration of O_2 carried in solution in the plasma. As an aside, the reduction of CO content in automobile exhausts by antipollution devices is apparently making of self destruction by CO poisoning more difficult.

Exercise

Humans can increase their O_2 consumption to many times the resting level during exertion. For most individuals, the transport capabilities of the ventilatory system are sufficient that the arterial blood remains completely saturated with O_2. In many cases, arterial P_{O_2} even increases during exertion, indicating the lung may be hyperventilated with regard to O_2. In these cases, there is clearly not a diffusion limitation to gas exchange. It is curious that many individuals are given gas mixtures with high P_{O_2} as an aid to recovery when there is no way that this diffusion-limitation based treatment can improve O_2 transport. If the Hb is already saturated, the problem is perfusion limitation. The only effect that higher P_{O_2} can have in these cases is to inhibit ventilatory drive and slow the heart rate. These effects would actually reduce O_2 transport!

Some humans do increase their perfusion enough during high levels of exercise that the arterial O_2 content falls. Many of these individuals are unaware of the transition to diffusion limitation of their gas exchange because their venous O_2 content falls with the arterial content and the $a-\bar{v}$ difference does not decrease. In this case, O_2 delivery is unimpaired, but the tissues must be able to function at the low P_{O_2} necessary to extract sufficient O_2 from the blood.

The immediate advantage of this response is not obvious. Perhaps it results in a lower ventilatory cost at high rates of oxygen consumptions. By allowing the alveolar P_{O_2} to fall the high cost of ventilation necessary to maintain a high P_{O_2} in the lung to prevent the diffusion limitation can be reduced. The limit on the extent to which desaturation can be useful will be a function of the lowest $P\bar{v}_{O_2}$ that can be tolerated. The mixed venous blood is just that, it combines blood from organs with lower rates of consumption with the blood from active muscles. The mixed venous O_2 content can never be zero. Even if the muscles were able to function with essentially no gradient for diffusion, the blood coming from other organs would not be totally desaturated. Perhaps the key to higher utilization of oxygen by active tissues lies in the ability to partition the blood flow away from less active tissues.

Some Thoughts About the Hb–O₂ Equilibrium Curve

The shape of the Hb–O_2 equilibrium curve appears to be adapted for loading O_2 at P_{O_2}'s that are well below what most mammals encounter at their gas exchange surfaces. One explanation for this conservative feature of gas transport is that is it an evolutionary relic from our aquatic vertebrate ancestry. Stagnation and hypoxia are two characteristics of water as a medium for gas exchange. Aquatic animals had to be able to extract O_2 from environments that were often hypoxic. But, why should this adaptation to hypoxia persist in air breathers? When vertebrates evolutionarily emerged from water into the O_2-rich medium of the air, their problems of hypoxic hypoxia were largely over, or were they?

An aquatic gas exchanger, like the gill, is supported by a medium of about the same density as the blood. In the lung, as we have discussed earlier, the difference in gravitational effects leads to much greater problems of inequities of ventilation and perfusion. In a region of the lung where the $\dot{V}/\dot{Q}$ ratio favors perfusion, the alveolar gases may become quite hypoxic. In this case, one can explain the persistence of the adaptation to hypoxia in air breathers. The inevitable $\dot{V}/\dot{Q}$ problems that arise from carrying on gas exchange between media of such different densities as air and blood provide a selective pressure to maintain blood that can extract O_2 from hypoxic regions of the lung. Perhaps the evolution of the lung could not have occurred if this preadaptation to $\dot{V}/\dot{Q}$ inequity were not left over from our gilled forbearers.

CHAPTER 10

RESPIRATORY TRANSPORT OF CARBON DIOXIDE

As you read about the transport of CO_2 in this and the remaining chapters, keep in mind the multiple functions of the respiratory system. Ventilation and circulation must maintain the removal of CO_2, the major end product of cellular metabolism, at a rate equal to its production. Simultaneously, the P_{CO_2} of the extracellular fluid compartments must be carefully regulated because of its importance to acid-base status. These two functions are obligately linked by their substrate and this linkage occasionally leads to a conflict between them. To understand this interaction, one must grasp the multiple relationships of CO_2 with its chemical partners.

Transport of CO_2 in the Blood

CO_2 behaves similarly to oxygen as it diffuses through the gases of the lungs and liquid of the interstitial fluid. Its transport in the blood, however, is much more complicated than oxygen. The ventilation of the lung determines that, because β_{gas} in air is the same for all gases and the amounts of CO_2 and O_2 simultaneously exchanged are nearly equal, the ΔP_{CO_2} in the lung in gas exchange must be the essentially the same as the ΔP_{O_2}. Every airbreathing animal that does not have an alternate mechanism for CO_2 removal has no choice in this matter because of the R-line considerations we discussed in Chapter 6 (refer to Fig. 6-2 and the discussion that begins on page 67 for a review of the topic). Therefore, the P_{CO_2} of the blood will be, at the least, as high above the atmospheric P_{CO_2} as the P_{O_2} of the blood is below its atmospheric level.

Within the blood, the β_{CO_2} is much greater than the β_{O_2}. The ΔP_{CO_2} between the arterial and venous blood will be, therefore, much smaller than the ΔP_{O_2}. This change must occur over and above the ΔP_{CO_2} between the blood and atmosphere we know must already exist for gaseous exchange in the lung. The minimum (that is, arterial) P_{CO_2} in the blood is equilibrated with the P_{CO_2} in the alveolus. The maximum (that is, venous) P_{CO_2} must be higher, so that CO_2 will diffuse into the alveolus.

In the case of humans, the minimum P_{CO_2} in the blood is normally about 5 kPa, representing a ΔP_{CO_2} of 5 kPa from the atmospheric P_{CO_2} of essentially 0 kPa. The alveolar ΔP_{O_2} is about the same, 5 kPa below the atmospheric level of 20 kPa. The corresponding changes of partial pressure of gases transported in the blood during this exchange are for CO_2, about 1 kPa, and for O_2, about 8 kPa. Note that the requirements of the aerial gas exchanger result in a residual amount of CO_2 remaining in the blood that substantially exceeds the amount exchanged in the lungs or tissue in any one pass through the exchanger.

The partial pressure of CO_2 with which the blood and other body fluids are equilibrated determines, in turn, the extent to which CO_2 combines with each of its various forms of transport. The amounts of CO_2 combined with these forms are, therefore, variables that are each *dependent* on the P_{CO_2}. The simplest of the forms in which CO_2 combines with all body fluids is solution in water. The solubility of CO_2 in specific fluids will be affected in part by the presence of other solutes but remains relatively constant at a given temperature.

Carbon dioxide interacts with water chemically in a series of reactions that produce first, carbonic acid and subsequently the dissociation products of this weak acid as shown below.

$$CO_2 + H_2O \rightleftharpoons H_2CO_3 \rightleftharpoons HCO_3^- + H^+ \rightleftharpoons CO_3^{2-} + H^+$$

In the electrochemical environment of the extracellular fluid, the bulk of the CO_2 involved in this reaction exists as bicarbonate, HCO_3^-, which is also the form of CO_2 that is found in the highest concentration in the blood. The concentration of the carbonate ion, CO_3^{2-}, is so small that its presence is usually neglected.

Carbon dioxide also interacts with the amino terminal ends of protein molecules. The blood differs from other extracellular fluids in that it has a much higher protein concentration. Not only are the plasma proteins more concentrated than in the interstitial fluid but also the red blood cells contain a still greater concentration almost entirely of the single protein, Hb. This concentration of Hb provides a substantial means to increase the carriage of CO_2 in the red blood cell directly. This form of carriage is given the name *carbamate* (or *carbamino*) binding of CO_2.

The sum of these CO_2 carriages, each of which contributes to the overall capacitance of the blood for CO_2, results in a concentration of about 22 mM/L (50 ml CO_2/100 ml) in the normal range of blood P_{CO_2}. For comparison, recall that the O_2 concentration of arterial blood is about 9 mM/L. The effect of the blood's high capacitance for CO_2 is that the concentration of CO_2 in the blood is higher than the concentration of O_2, even though the P_{CO_2} is less than the P_{O_2}. The apportionment of CO_2 carriage has been assigned different values by different authors, but a reasonable range would be CO_2 in solution, 7–10%, bicarbonate, 60–80%, and carbamate, 11–30%.

The blood–CO_2 dissociation curve given in Fig. 10-1 is the

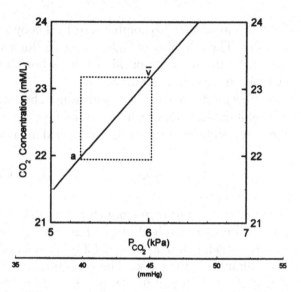

Fig. 10-1. CO_2 concentration and partial pressure in blood at the range of gas exchange.

result of the summation of all the forms of carriage of CO_2 in the blood. It has been adapted from Fig. 1-3 to emphasize the portion of the curve encompassing the normal arterial and venous concentrations and partial pressures. In contrast to the Hb–O_2 equilibrium curve, there is no plateau region in the capacitance of CO_2 in the blood. The importance of this difference is that CO_2 in the blood, despite the complexity of its carriage, behaves more like a simple, highly soluble, gas than O_2. Changes in concentration and partial pressure are essentially related directly within the normal physiological range. Put another way, the carriage of CO_2 in the blood, unlike oxygen, is not saturable at any normal P_{CO_2}. Any factor that affects the amount of CO_2 in the blood will be reflected by commensurate changes in both partial pressure and concentration.

 Carbon dioxide moves from the tissues into the blood by diffusion through the interstitial fluid, the capillary wall, and into the plasma. In the case of the molecules that associate with the red blood cell, the diffusion path must be extended to include the cell membrane and the intracellular fluid (ICF) of the red blood cell. The high capacitance of body fluids for CO_2 and its high solubility in cell membranes allow rapid equilibration of the P_{CO_2} along this diffusion path in all compartments as the blood flows between the tissues and the lungs.

Recall that the amount of O_2 carried in forms other than bound to Hb in the red blood cell is so small that we generally ignore it and treat O_2 transport as though it all occurred in that single compartment. In the case of CO_2 carriage, blood must be considered as physically having two compartments, the plasma and the red blood cell cytosol. The red blood cells comprise from 40% to 45% of the blood volume. In that volume, CO_2 exists in all three forms, notably the carbamate form which is, because of the high concentration of Hb, greater than the plasma ($[Hb]_{rbc}$ > 3 × $[total\ protein]_{plasma}$). Table 10-1 lists some typical values for venous and arterial concentrations of CO_2.

Of the total CO_2 transferred by the blood, the majority comes from bicarbonate and the majority of that from the plasma compartment (note from Table 10-1 that $\Delta[HCO_3^-]_{plasma} = 10 \times \Delta[HCO_3^-]_{rbc}$). At this point it is useful to look at the chemistry of the equilibrium between CO_2 and HCO_3^-. As is evident from Table 10-1, the equilibrium is such that, in the electrochemical

Table 10-1. Carriage of CO_2

Compartment	Forms of Carriage	Arterial Conc. (mM/L)	Venous Conc. (mM/L)	$\bar{v}-a$ Conc. Difference (mM/L)
Plasma	Dissolved CO_2	0.7	0.8	0.1
	HCO_3^-	15.2	16.2	1.0
	Carbamate	Trace	Trace	≈0
Red blood cells	Dissolved CO_2	0.3	0.4	0.1
	HCO_3^-	4.3	4.4	0.1
	Carbamate	1.0	1.4	0.4

Quantity (mM/L) exchanged from 1 L of blood assuming hematocrit = 45%

Plasma	0.6
Red blood cells	0.3

environment of the blood, the total $[HCO_3^-]$ is the most abundant form.

The relatively large total amount of HCO_3^- represents chemical equilibrium with the arterial P_{CO_2} and is residual in the blood. It is an amount that does not change under steady-state conditions of gas exchange. Comparatively small changes between the venous and arterial $[HCO_3^-]$ of the plasma are responsible for the majority of the gas exchange in that chemical form. In the red blood cell, although bicarbonate is the largest total constituent, the majority of the flux occurs as changes in the carbamate form. These changes too are smaller than the residual amount of CO_2 in that form.

The process by which CO_2 carriage in the blood is traditionally thought to occur is rather tortuous. The reaction between CO_2 and HCO_3^- is often described as the limiting step in CO_2 exchange because the uncatalyzed hydration reaction forming H_2CO_3 is too slow to allow gas exchange in the time in which it is observed to happen. The enzyme *carbonic anhydrase* (CA) catalyzes the reaction, and, in the blood, it is found in high concentration only in the red blood cell. The CA molecule is rather small, so if it were free in the plasma it would be removed by the kidney. The reaction between CO_2 and HCO_3^- in the plasma, which has been studied *in vitro*, that is, outside of the body, occurs very slowly without CA. The presence of CA in the red blood cell and the observation that between arterial and venous blood, there is a increase in the $[Cl^-]$ in the red blood cell has resulted in a complicated explanation for CO_2 transport in the blood called the *chloride shift*.

In the tissues, according to this explanation, CO_2 diffuses into the plasma and then into the red blood cell where CA rapidly converts it to HCO_3^-. The production of HCO_3^- ions then results in diffusion back into the plasma, causing all the change in $[HCO_3^-]$ in the plasma because the uncatalyzed reaction in the plasma would account for very little. To preserve electroneutrality, Cl^- is said to move back into the red cell as a counter ion to the HCO_3^-.

Cell membranes are relatively impermeable to ions. In this

case, the movement of anions across the red cell membrane appears to be enabled by the Band 3 protein that occurs abundantly in the membrane. It is assumed that the Cl^- and HCO_3^- ions diffuse simultaneously in opposite directions through this porter and result in the changes in concentration observed in the red blood cell and plasma. The whole process is then reversed as CO_2 leaves the blood in the lung. HCO_3^- diffuses back into the red blood cell and Cl^- back into the plasma. The CO_2 that reforms in the red blood cell then diffuses back into the plasma and then into the alveolus. Although this pathway is accepted and taught by the vast majority of physiologists and it is what students must learn to get "correct" scores on standardized exams, it has never made sense to me. What follows is what I think is a simpler and much more likely scheme (Prange and Sikora, 1995).

There is only one set of circumstances in which the pathway of the conventional model can make sense. First, the reaction in the red blood cells must be so much faster than the reaction in the plasma that that pathway is the only available pathway. Second, the reaction in the red blood cells must maintain a sufficient diffusion gradient for HCO_3^- for it to move in quantity between the cells and the plasma.

The reaction in the plasma, when studied outside the body, is indeed slow because there is no CA in the plasma. However, the presence of a membrane-bound form of CA in the capillary endothelium, unknown to those who studied the reaction *in vitro*, has now been shown. It can rapidly and sufficiently catalyze the exchange between CO_2 and HCO_3^- so that the reaction that mobilizes CO_2 from the plasma can keep up with its diffusion in or out of the blood (Heming et al., 1994). In its natural location, associated with the capillary endothelium, the plasma reaction can proceed rapidly. This discovery both obviates the need and eliminates the gradient for transit of HCO_3^- through the red blood cell. If we approach the transport from this perspective, many of the other phenomena of CO_2 transport in the blood fall into place.

If no diffusion gradient for bicarbonate exists, then we must account for the movement of Cl^- ions. The movement of Cl^- through its porter will occur under only two conditions. First, if there is a concentration gradient of Cl^- across the membrane and/or second, there is an electrical potential gradient to which it can respond. Another phenomenon of gas exchange, also unknown to the formulators of the conventional hyppothesis, is that the Hb molecule binds reversibly with Cl^- ions as a function of its oxygenation. As oxygen leaves Hb and diffuses to the tissues, Cl^- ions are bound to it, decreasing the $[Cl^-]$ in the red blood cell and creating a gradient for diffusion into the cell from the plasma. This change in $[Cl^-]$ is critical to the increase in $[HCO_3^-]$ and to the acid-base balance in that compartment.

As I will explain at greater length in Chapter 11, the electrochemical environment of the red blood cell cytosol, as evidenced by changes in $[H^+]$, is strongly affected by the independent variables, P_{CO_2} and $[Cl^-]$. The red blood cell compensates for the effects wrought on its cytosol by the changing P_{CO_2} with the only other independent variable available, the difference in concentration of strong ions, in this case Cl^-. The chloride shift occurs between the plasma and the red blood cell in response to and simultaneously with the movement of CO_2 in and out of the cell, but it has nothing to do with any opposite movement of HCO_3^-.

The amount of CO_2 that is transported as HCO_3^- in the red blood cell is a small but important component of overall CO_2 transport. Inhibition of this reaction may have little effect on CO_2 transport at rest, but it does have an effect at higher rates of metabolic activity. The CA in the red cell is important to the movement of CO_2 in and out of the cell which must occur rapidly during the brief transit time through the exchange capillaries. However, the CA in the cytosol of the red blood cell has nothing to do with the $[HCO_3^-]$ in the plasma.

Rather than a serial path through the red blood cell required by the chloride shift hypothesis, we can now envision the transport of CO_2 in the blood as a path that splits within the plasma, as is shown in Fig. 10-2 (Prange and Sikora, 1995). The majority of the CO_2 is in rapid equilibrium with the HCO_3^-. A smaller

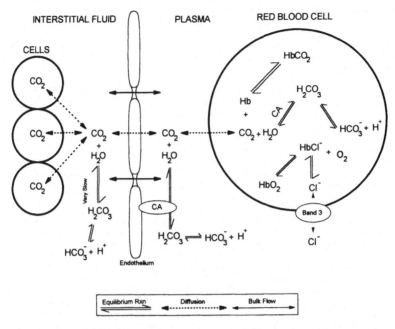

Fig. 10-2.   The pathway for transport and exchange of CO_2 between the tissues and lungs. The mechanisms are reversible, so either the lungs or the tissue may be represented as "cells."

part of the CO_2 diffuses in and out of the red blood cell and is in rapid equilibrium there with the carbamate and HCO_3^- modes of transport. Cl^- and water move passively through their respective porters to maintain homeostasis of the electrochemical environment and osmotic equilibrium.

CHAPTER 11

RESPIRATORY CONTROL OF ACID–BASE

The topic of acid–base control as it is traditionally presented is difficult to understand because of awkward conventions, such as the use of pH to represent concentration of H^+ ions, and because of a misunderstanding of the dependent and independent variables and the nature of the equilibria in which they are involved. In fact, what is meant by control of acid–base is the control of the electrochemical environment that, in turn, affects the structure and function of the many types of molecules that are in it. Throughout this text I have referred to the electrochemical environment only as the milieu of the chemistry of the extracellular fluid. In this chapter, I hope to make it clear what that environment is and how it works. As it turns out, it really is much simpler than we have been led to believe.

Neutrality

Pure water (HOH) chemically dissociates to a slight extent into hydroxyl ions (OH^-) and hydrogen ions (H^+). At 25°C, the dissociation constant of water (K_w') is 10^{-14} if we assume the concentration of the undissociated HOH (55.3 M) does not measurably change. The concentrations of OH^- and H^+ are, therefore, each 10^{-7} molar and their ratio is 1:1. This ratio defines neutrality and the sense of acid-base. Although the equilibrium constant changes with temperature so that, at 37°C it is about $10^{-13.6}$ and at 0°C it is about 10^{-15}, acid-base neutrality is always defined by the equality of the OH^- and H^+ concentrations, whatever their actual values may be.

What is pH, Anyway?

Concentrations so small that they must be written with negative exponents are easily misplaced, so it may be convenient to use a system with simpler and larger numbers. If we represent the $[H^+]$ as $1/\log[H^+]$, it has the value of 7.0 at neutrality and 25°C. This representation, called pH in this case, can also be used for other similarly small values; the $[OH^-]$ has a pOH of 7.0 under the same conditions and K_w' has a $pK_w' = 14$. Describing things as the inverse of their logarithm may seem a bit odd in that not only does the pH value change in opposite magnitude to the $[H^+]$ it represents, but also a change in pH one unit larger is not the same difference in concentration as a change one unit smaller. Put mildly, the conversion of pH values back to the actual concentrations is anything but intuitively obvious. There is (or was, at least) another justification for the use of pH units (Text Box 11-1). As an additional point, note that because of the temperature dependence of the K_w', neutral pH is 7 only at 25°C, roughly room temperature. This value for neutrality is what we all learned in chemistry class. At human body temperature, 37°C, neutral pH is 6.8.

Relative Alkalinity of the Extracellular Fluid

The normal pH of the extracellular fluid (ECF) is about 7.4, or, put another way, 0.6 pH units on the alkaline side of neutrality.

Text Box 11-1. The Origin of pH

The development long ago of a form of glass that was selectively permeable to hydrogen ions made it possible to measure the electrical potential that was generated by the difference in $[H^+]$ between a solution and the interior of an electrode. The electrical potential between the electrode enclosed in the glass and one placed in the solution can be described by the Nernst equation as a logarithmic function of the concentration difference.

Electrical potential was, at this time measured directly from sensitive moving-coil galvanometers. It was difficult to read the actual concentration on the scale of these meters because it was logarithmic, so conversion was made to a simple linear scale, accomplished by use of the inverse log convention and, *voilà*, we have pH.

Now, with digital conversions of voltages to logarithms that are possible with modern methods, the need for obscuring the $[H^+]$ with the use of pH no longer exists, but tradition dictates that this old-fashioned convention persist.

Translating this number back into concentrations, we can see that it means the normal $[H^+]$ is about 40 nanomoles per liter and that the ratio of $[OH^-]$ to $[H^+]$, from the difference of pOH and pH, is about 30:1. This ratio appears to describe the appropriate acid–base status for the electrochemical environment for vertebrates because many species maintain it irrespective of the temperature or actual $[H^+]$ or pH of their ECF (Howell et al., 1970). Whatever other changes may occur, the molecules that depend on the electrochemical status of this environment for their proper function will find conditions satisfactory if the ratio of about 30 OH^- ions to every H^+ ion is maintained.

Dependent and Independent Variables

We often speak casually and causally, but entirely incorrectly, of the electrochemical environment being controlled by pH. It is important to clarify and emphasize the point that the *concentra-*

tion of H^+ ions itself in no way controls the electrochemical environment. The concentration of hydrogen ions, like other dependent variables in the electrochemical environment, only indicates the status of that environment by its concentration; it does not determine it. Nothing about the structure of a given enzyme or other protein is controlled by the $[H^+]$ around it. Both the structure of a protein and the $[H^+]$ are dependent on the independent variables, those variables that can be manipulated directly in the system.

The nature of the dependent and independent variables in acid–base has been clearly described by Stewart (1980) and has been related to more applied and clinical settings by others, for example, Fencl and Rossing (1989) and Fencl and Leith (1993). It is not my purpose here to give the fundamentals of this concept a full review but rather to show how they apply to respiratory effects on the electrochemical status of the ECF.

Like all chemical solutions, the ECF is not exempt from adherence to fundamental chemical and physical principles. Mass is conserved, electroneutrality of the fluid is maintained, and the equilibrium constants for the dissociable species operate in the body as they would in a test tube. We need only to consider which components of the ECF fall into the categories of dependent and independent variables and to keep these fundamental principles in mind to understand the control of the system. The regulatory systems can alter the system only by directly altering the independent variables, which we will see are only three, P_{CO_2}, with which we are familiar, and the total weak acid anion concentration and strong ion difference, to be defined below. The hydration reaction of CO_2 and water and the subsequent dissociation into HCO_3^- and its other products occurs under the control of the only independent variable in the reaction, P_{CO_2}.

The changes in the independent variables will cause changes of the dependent variables; the opposite is never the case. So, when we speak casually of pH causing some change, what we really mean is that the independent variables, some of which may be difficult to measure directly, have caused pH, which we can observe easily, and the other dependent variables to change.

If we consider the strong electrolytes of the ECF, such as Na^+, K^+, and Cl^-, we see that, at physiological concentrations, these species are fully dissociated. No changes in the concentration of other constituents of the ECF can alter their concentration. Only the direct physical addition or removal of these ions from the system can effect that change.

In contrast to these strong electrolytes, weak electrolytes, that is, weak acids and bases, are characterized by transitions between their associated and dissociated forms as a function of changes in other constituents of the solution. The concentrations of the dissociation products of weak acids can change and do, according to the dictates of the law of mass action. In the example below, the anion portion of the weak acid is given the general symbol A^-:

$$HA \leftrightharpoons H^+ + A^-$$

The H^+ ions of this equilibrium participate in all of the other weak acid dissociations in the ECF. They do not simply belong to a particular weak acid in question so their concentration may change as a function of other reactions. The interrelatedness of all the weak acid dissociations allows one to consider them as though all were governed by a common dissociation constant.

The total concentration of the weak acid anion (A^-) in its dissociated and combined forms does not change unless that substance is itself consumed or added. No matter what happens to the $[H^+]$, the sum of their concentration remains constant. By our previous definition, the sum of the forms of the anion is, therefore, also an independent variable. That sum is called *total acid anion concentration* ($[A_{tot}]$). The equation below repeats this definition:

$$[A_{tot}] = [A^-] + [HA]$$

The effects of the independent strong electrolytes can be collected into a single independent variable, the *strong ion difference* ($[SID]$). $[SID]$ is defined as the difference between the total concentration of strong cations and the total concentration of all the strong anions. With that, the list of independent variables is complete; there are only three: P_{CO_2}, $[A_{tot}]$ and $[SID]$.

How Acid–Base Is Regulated

All of the other constituents of the system (the dependent variables) can be represented in relationships derived from the fundamental principles mentioned above as functions of these three independent variables. The *only* means by which the body can internally alter the status of the electrochemical environment is by manipulation of one of the three independent variables. The *only* way we can exert any external correction on an individual's electrochemical status is by manipulation of one of the three independent variables. It is that simple.

Now it will get even simpler. The $[A_{tot}]$ in a fluid compartment tends to vary little and slowly. $[A_{tot}]$ consists largely of proteins and small amounts of weak acids. The concentrations of proteins in the blood are largely controlled by the liver's synthesis and degradation processes. In other fluid compartments, $[A_{tot}]$ results from the rather steady output and consumption of the net metabolism of the cells. $[A_{tot}]$ is important to the establishment of the status of the electrochemical environment, but, except in the very long term, the $[A_{tot}]$ is not an active player in its regulation.

The important variables in ongoing acid–base regulation are therefore only two: P_{CO_2} and $[SID]$. The ventilatory system regulates the P_{CO_2} and the kidney regulates $[SID]$, and that's that. Those of you who have been subjected to the conventional treatment of acid base and may be concerned that I have made no mention of buffers, the Henderson–Hasselbalch equation, or pH–bicarbonate diagrams, read on.

Buffers work best when the pH of their solutions is near their pK. This observation has two important ramifications. First, only those buffers with a pK near 7.4 will be important to the ECF.

HCO₃, the weak acid anion is by far the highest concentration in the ECF, has a pK of 6.1 that is well removed from 7.4, so its buffering effect is not important. Second, as Stewart has elegantly pointed out, buffers buffer pH, not the electrochemical environment. Because $[H^+]$ is a dependent variable, the only importance of the so-called buffers is as they contribute to $[A_{tot}]$ and that is a static contribution.

The importance of the buffering effect of Hb on the blood acid-base status is often invoked. Hb contributes importantly to the $[A_{tot}]$ of each red blood cell, but its effect on acid-base is only that. Hb is important to the electrochemical environment of each cell, but it, like other intracellular constituents, has no importance to the acid–base status of the ECF.

The Henderson–Hasselbalch equation (Hasselbalch, 1916) describes the relationship of two dependent variables, pH and $[HCO_3^-]$, two constants, pK (= 6.1 at 37°C) and $\beta_{blood_{CO_2}}$ (recall that β is the capacitance coefficient, in this case, of blood for CO_2), and an independent variable, P_{CO_2}. For convenience and recognition, I will give it below.

$$pH = 6.1 + \log\left(\frac{[HCO_3^-]}{P_{CO_2} \times \beta_{CO_2}}\right)$$

The relationship is true and the variables are measurable. But, it leads to the mistaken notion that the body can and does control its acid–base status by manipulation of the plasma $[HCO_3^-]$. Further, two important independent variables, [SID] and $[A_{tot}]$ are omitted from the Henderson–Hasselbalch equation and, thus, from consideration by those trying to solve the complexities of the control of the electrochemical environment of the ECF. The virtue of the Henderson–Hasselbalch equation is that its variables are easily measured. [SID] and $[A_{tot}]$, because they are sums of concentrations of several components are difficult, if not impossible, to measure directly. So long as we consider $[HCO_3^-]$ and pH only as indicators and not as controllers, we will not misunderstand what is really going on (see Text Box 11-2).

Text Box 11-2.   The Usefulness of pH and [HCO$_3^-$]

I like to think of pH and [HCO$_3^-$] with the analogy of flags blowing in the wind. We observe the direction of the wind readily from the position of the flag because it is much easier to see the flag than the wind that moves it. Similarly, we observe the effects of the independent variables in acid–base, P_{CO_2}, [SID] and [A$_{tot}$], by observing pH and [HCO$_3^-$].

We would never suggest that if we could move the flag, the direction of the wind could be changed. Neither should we suggest that we could change the status of the electrochemical environment if we could alter only its indicators, the dependent variables such as pH or [HCO$_3^-$].

The pH–bicarbonate diagram, aka, "Davenport diagram" after its source [see Davenport (1969)], for example, is a graphical representation of the Henderson–Hasselbalch equation. It presents the regulation of acid–base as a function of the two dependent variables in the equation: pH and [HCO$_3^-$]. P_{CO_2}, the only independent variable, is represented in isobars, as though it were a constant. Thus we are mislead in three different ways by the presentation.

A much more reasonable format for understanding would be to use a graph with independent variables, P_{CO_2} and [SID] (Fig. 11-1). The pH-bicarbonate diagram, because it employs the convenient indicators on it axes and because its use is so entrenched in tradition, is useful for diagnosis of acid-base problems. Its use should be discouraged if we are to understand the nature of control of the electrochemical environment.

Respiratory Control of the Electrochemical Environment

Ventilation may exceed the gas flow necessary to transport the CO_2 production from the body. Irrespective of the actual $\dot{V}$, this

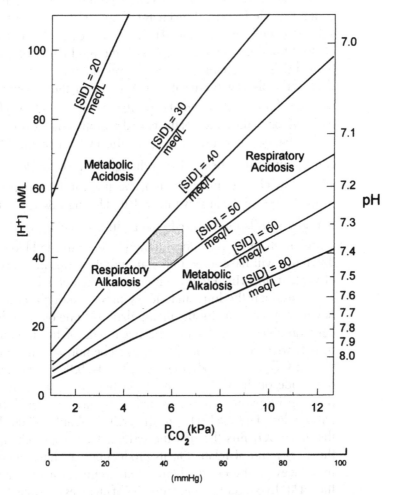

Figure 11-1. Acid-base of plasma as a function of [SID] and P_{CO_2} after Stewart, (1980). The shaded area represents the approximate normal range for humans. Compensation for changes in one independent variable takes place via changes in the other. For example, if hyperventilation resulted in decreased P_{CO_2}, the kidneys would decrease the [SID] to return the electrochemical environment to normal. The effect of this compensation would be manifested in a return of [H$^+$] toward the normal 40 nM/L.

circumstance is considered *hyperventilation* because the CO_2 content of the blood will decrease. Conversely, *hypoventilation* will allow the CO_2 in the blood to increase because the ventilation of the lung does not keep pace with the production of CO_2 by the cells. These changes in relative ventilation can control the P_{CO_2} of the blood without substantially altering the oxygen supply because of the much different capacitances for O_2 and CO_2 found in the blood in the region of the partial pressure changes of normal gas exchange. The low $\beta_{blood_{O_2}}$ in the plateau of the Hb–O_2 equilibrium curve means that, although the alveolar P_{O_2} may be affected by changes in ventilation, the amount of O_2 transported by the blood will be largely unaffected. The high $\beta_{blood_{CO_2}}$ means that any ventilation-induced change in alveolar P_{CO_2} will have a similar effect on the concentration of CO_2 in the blood. This difference will have some important ramifications in the control of ventilation, which will be discussed in Chapter 12.

Although the lung and kidney may be thought of as each having responsibility for the manipulation of one of the independent variables in acid–base, the magnitude of the responses is heavily weighted in favor of the lung. The normal daily production of CO_2 in the body may be 20,000–30,000 mM. On the other side of the CO_2 hydration reaction, the total amount of $[H^+]$ in the plasma and interstitial fluid may be a few hundred nanomoles. The total $[H^+]$ in the ECF is about 0.000001% of the daily CO_2 flux through the system. Even slight changes in the CO_2 removal relative to its production can have enormous influences on the electrochemical environment as indicated by the $[H^+]$. In contrast, the magnitudes of changes possible through the kidneys' effects on [SID] are much smaller and slower. The kidney can only effect transport that would give the equivalent of about 750 mEq of H^+ per day.

When the ventilatory transport of CO_2 exceeds the CO_2 production so that the P_{CO_2} declines the effect is seen in a decreased $[H^+]$ and is termed *respiratory alkalosis*. Recall that the $[H^+]$ is normally on the alkaline side of neutral to maintain the 30:1 ratio of $[OH^-]$ to $[H^+]$. In respiratory alkalosis, that ratio is increased; the ECF becomes more alkaline. An increase in the $[OH^-]/[H^+]$ ratio, not the change in pH, should be the criterion for alkalosis. Respiratory alkalosis can occur as a response to the low P_{O_2} of high altitude or as a consequence of nervous hyperventilation. In pneumonia, the fluid in the lungs impedes CO_2 transport much less than O_2 because of the high capacitance of water for CO_2. Ventilation that is increased to maintain a high ΔP_{O_2} across the diffusion barrier will result in excess removal of CO_2 and consequent respiratory alkalosis.

Respiratory acidosis occurs when the ventilation is less than that required to remove the CO_2 produced and the $[H^+]$ consequently increases. Anything that suppresses ventilatory drive such as depressants of the central nervous system or impedes normal ventilation such as obesity or pneumothorax can lead to respiratory acidosis.

In chronic cases of respiratory acid–base imbalance, renal handling of [SID] will compensate, albeit rather slowly in an animal as large as a human, perhaps in hours or even days, to the extent the transport capabilities of the kidney allow. The large amount of CO_2 handled by the lung allows respiratory compensation for acid–base disturbances to be much more rapid.

There is controversy over the use of the so-called *strong ion difference* approach to acid–base. [SID] is not conveniently measurable and, for purposes of diagnosis, the measurement of pH and P_{CO_2}, and subsequent calculation of $[HCO_3^-]$, are correct and convenient methods. What is irrefutable is the correctness of the basic chemistry that [SID] discloses and its importance to understanding the physiological mechanisms by which the body regulates the status of the electrochemical environment. Explanations based on treatment of $[H^+]$ (or pH) and $[HCO_3^-]$ as independent variables, including their transport across membranes, are wrong and misleading and should be consigned to the realm of fantasy.

CHAPTER 12

CONTROL OF

VENTILATION

Ventilation serves many masters. I have emphasized ventilation as it serves gas exchange and acid–base regulation, but we should not forget its important, if less obvious functions, such as vocalization, sneezing, coughing, cooling hot coffee, and, in some species, evaporative cooling of the body. All these functions are somehow superimposed and prioritized in a combination of voluntary and autonomic control that is understood more in terms of cause and effect than in the nature of the actual control mechanisms.

Rhythmicity

The normal tidal breathing of an individual is obviously rhythmic when other demands on ventilation are not imposed. It is useful to compare ventilation of the lungs with that other rhythmic pump physiologists concentrate on, the heart. The heart tends to fill passively and empties forcefully, the opposite of ventilation of the human lung at rest. The autonomic innervation of the heart can alter its rhythm but, unlike ventilatory muscles, the cardiac muscle will continue to contract rhythmically in the absence of innervation.

The muscles of ventilation are skeletal and entirely dependent on innervation for contraction. The phrenic nerve controls the movements of the diaphragm and the intercostal muscles that move the ribs are innervated by spinal nerves. In contrast to the heartbeat, the rhythmic pattern of ventilation originates entirely in the brain. The possible origins of these patterns will be discussed below.

Chemoreceptors

Normal breathing, composed of regular relatively shallow breaths, can be adjusted on a breath-by-breath basis. The most apparent inputs to the control of ventilation are the peripheral and central *chemoreceptors* that respond to changes in the partial pressures of respiratory gases and acid–base status of the blood. In humans, the most important, though not the only, chemoreceptors for P_{O_2} are tissues in the carotid arteries called the carotid bodies. In other species, tissues elsewhere in the arterial circulation may take precedence.

P_{CO_2} chemoreceptors are primarily central and are located in the medullary part of the brain stem. In some cases, the responses of the ventilation appear to be compounded by stimuli from both kinds of receptors. The responses of human ventilation to changes in the arterial P_{CO_2} are the most sensitive and apparent. As can be seen in Fig. 12-1, for a given individual the relationship of ventilation to P_{CO_2} at normal P_{O_2} is essentially linear and steep.

It is difficult to manipulate arterial P_{CO_2} directly in human subjects, so most data are gathered from experiments in which alveolar P_{CO_2} is controlled with breathing mixtures having higher-than-atmospheric P_{CO_2}. Normal CO_2 exchange must continue via diffusion from the blood, so the arterial P_{CO_2}'s can be expected to rise to be higher than alveolar values until a sufficient gradient is created to maintain gas exchange.

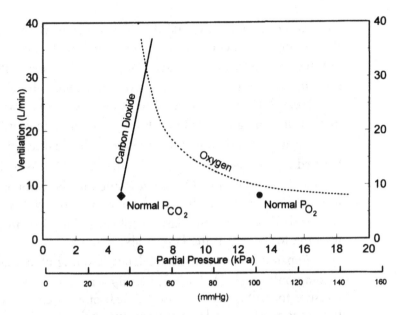

Fig. 12-1. Ventilatory responses to P_{O_2} and P_{CO_2}. Each curve represents experimental data from a single, different individual. Data from other individuals would almost certainly differ in the actual numbers but would illustrate the same general pattern of responses. For reference a symbol is included in the figure for the normal values of the partial pressures of the respiratory gases and a typical value for ventilation that might be found. The figure is based on data from Loeschke and Gertz (1958) and Nielsen and Smith (1952).

At best, one can get a qualitative understanding of the effects of arterial P_{CO_2} from these experiments on ventilation because the stimulus is unusual and the response to it, increased ventilation, does nothing to ameliorate the condition of higher-than-normal levels of CO_2 in the blood. Individuals' responses to CO_2 are highly variable and most data are presented as statistically-derived lines that obscure the extent of the variation in the subjects studied. Despite these constraints, it is reasonable to assume that the normal response to increased arterial P_{CO_2}, increased ventilation with CO_2-free air, can be generally modeled from these data.

Lowering of inspired P_{O_2} or reduced arterial P_{O_2} are phenomena that occur normally, and the relationship of ventilation to P_{O_2} is more easily modeled in laboratory experiments. The ventilatory response to decreased arterial P_{O_2} is qualitatively different from the response to CO_2. In a given individual, the alveolar P_{O_2} (Fig. 12-1) may be lowered to as little as 7–8 kPa with no increase in ventilation. At lower P_{O_2}'s the response of ventilation to P_{O_2} is more nearly linear. If anything, individual responses to lowered P_{O_2} are even more variable than to CO_2. As with the ventilatory response to CO_2, the response may be compounded by other influences. At low P_{O_2}'s the sensitivity to high P_{CO_2} may be doubled or more. Similarly, ventilation increases more in response to low P_{O_2} if the P_{CO_2} is above normal.

Apparent insensitivity to moderate lowering of arterial P_{O_2} is commonly observed. The relationship of arterial oxygen partial pressure to ventilatory response has led some to conclude that the P_{O_2} receptors are relatively insensitive or that P_{O_2} is unimportant to regulation because, at least in the case of hypoventilation, a sufficient stimulus to cause a response would be reached with an elevated P_{CO_2} long before the P_{O_2} would have fallen to a threshold level.

With a more careful examination of the relationship of partial pressure to ventilation, one can see that the response of ventilation to P_{O_2} is both sensitive and appropriate. In Fig. 12-2 the ventilatory response curve and the O_2 equilibrium curve have been plotted together with the axes scaled appropriately. Now it should be apparent that the ventilatory drive corresponds to the reduction of O_2 saturation rather than to reduction of partial pressure. Of course, only a variable with intensive properties, like P_{O_2}, and not saturation, can be detected. The receptors have apparently evolved a nonlinear response that effectively matches an increase of ventilation to compensate for the decrease in saturation. Rather than being insensitive, the chemoreceptors are exquisitely matched to the variable they detect. They work to prevent unneeded hyperventilation when, even though the P_{O_2} may be lowered, the blood is still well saturated with O_2.

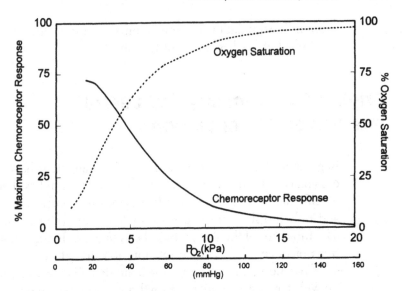

Fig. 12-2.   Comparison of the activity of a carotid chemoreceptor and the saturation of blood with oxygen as functions of P_{O_2}.

Mechanoreceptors

Vagal stretch receptors in the airway walls give information on the degree of inflation of the lung. As the lung becomes increasingly inflated, the output from these receptors has an inhibitory effect on inspiration. This reflex may operate to terminate very deep breaths but does not appear to cause the cessation of normal tidal inhalation.

Other Inputs to Ventilatory Control

Reflexive actions that interrupt ventilation such as coughing sneezing and swallowing are triggered by irritant or touch sensors in the pharynx and airways. Proprioceptors in joints and muscles have been implicated as stimulating increased ventilation during exercise and stretch receptors in the intercostal muscles may also play a part in regulation. Finally, higher centers can override the

autonomic stimulation of ventilation for some of the activities
mentioned at the beginning of this chapter.

Origin of Rhythmicity and Central Integration of Ventilation

Regions of the pons and medulla in the brain stem have been
designated *respiratory centers*. The term *center* implies that the
specific nature and function of these areas are poorly understood
and only grossly correlated with anatomically distinct features.
The functions of these areas have been inferred through observations of direct electrical stimulation and the effects of lesions,
serial ablations, and injuries of the brain stem. These observations
notwithstanding, the origin of rhythmic ventilation and the inte-
gration of inputs can be only partly explained.

During inhalation, the electrical activity of the inspiratory
neurons that send impulses to inspiratory muscles increases. It
wanes as expiration begins and resumes with the start of the next
ventilatory cycle. In humans, although not necessarily so in other
species, exhalation is passive at rest. Exhalation, whether at rest
or forceful as during exercise, is accompanied by increased activity
in the expiratory neurons. The rhythmicity may be from as un-
complicated an interaction as reciprocal inhibition by inspiratory
and expiratory neurons, but it probably is not that simple.

In the evolution of the vertebrate brain, functions that were
once under the control of the brain stem and lower areas are
successively assumed by higher centers. In many cases, removal
of or injury to the higher centers leaves a hierarchy of forms of
crude respiratory control. These forms are useful in the diagnosis
of injuries to the brain, but it is not clear whether they are still
an important part of ventilatory control or are evolutionary relics.

At best, we can summarize the control of ventilation with
a block diagram like Fig. 12-3. The most readily observed influences on ventilatory control are arterial P_{O_2} and P_{CO_2} and the
other variables that affect the electrochemical environment. Not

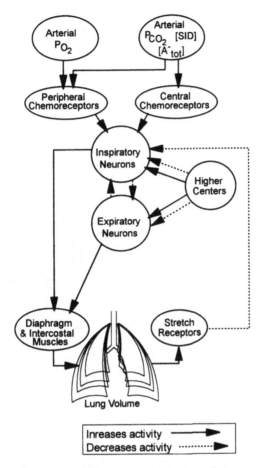

Fig. 12-3.   Scheme for control of ventilation.

all of the indicated communications operate in all circumstances. For example, the inhibition of stretch receptors may occur only in the case of deep inspiration. During exercise, arterial P_{CO_2} may actually be lower, and P_{O_2} higher, than at rest, so there must be other factors that match ventilation to the need for gas exchange.

The nature of the origin of rhythmicity, the drive of ventilation during exercise, and other pieces of the puzzle are still to be found and put in place.

BIBLIOGRAPHY

BACHOFEN, H., J. HILDEBRANDT, and M. BACHOFEN (1970) Pressure-volume curves of air- and liquid-filled excised lungs—surface tension in situ. *J. Appl. Physiol.* 29:422–431.

BOHR, C., K. HASSELBALCH, and A. KROGH (1904) Ueber einen biologischer Beziehung wichtigen Einfluss, den die Kohlensäurespannung des Blutes auf den Sauerstoffbindung übt. *Skand. Arch. Physiol.* 16:402–412.

BOYNTON, W. P. and W. H. BRATTAIN (1929) Interdiffusion of gases and vapors. In: *International Critical Tables of Numerical Data, Physics, Chemistry and Technology*, Vol. V, pp. 62–65. Washburn, E. W. (ed). McGraw-Hill, New York.

BUDYKO, M. I., A. B. RONOV, and A. L. YANSHIN (1987) *The History of the Earth's Atmosphere*. Springer-Verlag, Berlin.

CHRISTIANSEN, J., C. G. DOUGLAS, and J. S. HALDANE (1914) The absorption of carbon dioxide by human blood. *J. Physiol.* 48:244–277.

CUMMING, G., J. CRANK, K. HORSFIELD, and I. PARKER (1966) Gaseous diffusion in the airways of the human lung. *Respir. Physiol.* 1:58–74.

DAVENPORT, H. W. (1969) *The ABC of Acid–Base Chemistry*. 5th ed. University of Chicago Press, Chicago, IL.

FENCL, V. and D. E. LEITH (1993) Stewart's quantitative acid–base chemistry: applications in biology and medicine. *Respir. Physiol.* 91:1–16.

FENCL, V. and T. H. ROSSING (1989) Acid–base disorders in critical care medicine. *Ann. Rev. Med.* 40:17–29.

FENN, W. O. (1927) The oxygen consumption of frog nerve during stimulation. *J. Gen. Physiol.* 10:767–779.

FORSTER, R. E. (1964) Diffusion of gases. In: *Handbook of Physiology*, Vol. 1, pp. 839–872. Fenn, W. O. and H. Rahn (eds.). Waverly Press, Baltimore, MD.

GLAISTER, D. H., R. C. SCHROTER, M. F. SUDLOW, and J. MILIC-EMILI (1973) Bulk elastic properties of excised lungs and the effect of transpulmonary gradient. *Respir. Physiol.* 17:347–364.

HARVEY, E. N. (1928) The oxygen consumption of luminescent bacteria. *J. Gen. Physiol.* 11:469–475.

HASSELBALCH, K. A. (1916) Die Berechnung der Wasserstoffzahl des Blutes aus der freien und gebundenen Kohlensäure desselben, und sie Saurestoffbindung des Blutes als Funktion der Wasserstoffzahl. *Biochem. Z.* 78:112–144.

HEMING, T. A., E. K. STABENAU, C. G. VANOYE, H. MOGHADASI, and A. BIDANI (1994) Roles of intra- and extracellular carbonic anhydrase in alveolo-capillary CO_2 equilibrium. *J. Appl. Physiol.* 77(2): 697–705.

HENDERSON, R., K. HORSFIELD and G. CUMMING. (1968/69) Intersegmental collateral ventilation in the human lung. *Respir. Physiol.* 6:128–134.

HORSFIELD, K. and G. CUMMING (1968) Morphology of the bronchial tree in man. *J. Appl. Physiol.* 24(3):373–383.

HOWELL, B. J., F. W. BAUMGARTNER, K. BONDI, and H. RAHN (1970) Acid–base balance in cold-blooded vertebrates as a function of body temperature. *Am. J. Physiol.* 218:600–606.

IMAI, K. (1982) *Allosteric Effects in Haemoglobin.* Cambridge University Press, London.

KROGH, A. (1919) The rate of diffusion of gases through animal tissues, with some remarks on the coefficient of invasion. *J. Physiol.* 52:391–408.

LOESCHCKE, H. H., and K. H. GERTZ (1958) Einfluß des O₂-Druckes in der Einatmungsluft auf die Atemtätigkeit des Menschen, geprüft unter Konstantaltung des alveolaren CO₂-Druckes. *Pflügers Archiv.* 267:460–477.

METCALFE, J., H. BARTELS, and W. MOLL (1967) Gas exchange in the pregnant uterus. *Physiol. Rev.* 47:782–838.

NIELSEN, M. and H. SMITH (1952) Studies on the regulation of respiration in acute hypoxia. *Acta Physiol. Scand.* 24:293–313.

NEWHOUSE, M., J. SANCHIS, and J. BIENENSTOCK (1976) Lung defense mechanisms (first of two parts). *N. Engl. J. Med.* 295:990–998.

PIIPER, J., P. DEJOURS, P. HAAB, and H. RAHN (1971) Concepts and basic quantities in gas exchange physiology. *Respir. Physiol.* 13:292–304.

PIIPER, J and P. SCHEID (1972) Maximum gas transfer efficacy of models for fish gills, avian lungs and mammalian lungs. *Respir. Physiol.* 14:115–124.

PRANGE, H. D. and J. G. SIKORA (1995) Transport of carbon-dioxide in the blood: an updated model. *FASEB J.* 9 (3): A148.

RAHN, H., A. B. OTIS, L. E. CHADWICK, and W. O. FENN (1946) The pressure–volume diagram of the thorax and lung. *Am. J. Physiol.* 146:161–178.

ROSENBERG, E., P. ERNST, J. LEECH, and M. R. BECKLAKE (1986) Specific diffusing capacity (D_L/V_A) as a measure of the lung's characteristics: prediction formulas for young adults. *Lung* 164:207–215.

ROUGHTON, F. J. W. (1964) Transport of oxygen and carbon dioxide. In: *Handbook of Physiology*, Vol. 1, pp. 767–825. Fenn, W. O. and H. Rahn (eds.). Waverly Press, Baltimore, MD.

SCHMIDT-NIELSEN, K. (1972) *How Animals Work*. Cambridge University Press, London.

STEWART, P. A. (1980) *How to Understand Acid–Base: A Quantitative Acid–Base Primer for Biology and Medicine*. Elsevier, New York.

TORRANCE, J. D., C. LENFANT, J. CRUZ, and E. MARTICORENA (1970/71) Oxygen transport mechanisms in residents at high altitude. *Respir. Physiol.* 11:1–15.

VOET, D. and J. G. VOET (1990) *Biochemistry*, Wiley, New York.

WEIBEL, E. R. (1963) *Morphometry of the Human Lung*, Springer-Verlag, Berlin.

INDEX